Tropical Mycorrhiza Research

Tropical Mycorrhiza Research

Edited by
PEITSA MIKOLA

CLARENDON PRESS OXFORD 1980

Oxford University Press, Walton Street, Oxford OX2 6DP

OXFORD LONDON GLASGOW
NEW YORK TORONTO MELBOURNE WELLINGTON
KUALA LUMPUR SINGAPORE JAKARTA HONG KONG TOKYO
DELHI BOMBAY CALCUTTA MADRAS KARACHI
NAIROBI DAR ES SALAAM CAPE TOWN

Published in United States by Oxford University Press,
New York

British Library Cataloguing in Publication Data
Tropical mycorrhiza research.
   1. Mycorhiza
   2. Tropical plants
   I. Mikola, Peitsa
   589′.2′04524   QK604   79-42778
   ISBN 0-19-854553-3

Set by Hope Services, Abingdon
Printed by J. W. Arrowsmith Ltd., Bristol

Foreword

In 1973 the International Foundation for Science (IFS) instituted a programme of grants for research in developing countries. The five subject-areas, in support of agriculture in the broadest sense, included research on mycorrhiza which was considered to be highly relevant. By 1978 the Foundation had approved only twenty grants for mycorrhiza research, a modest proportion (8 per cent) of the total grants.

The relatively few institutions receiving grants are widely separated in south-east Asia, Africa, and Latin America. For this reason they find it difficult or impossible to confer among themselves and with other experts in the field. Perceiving this, the IFS decided to arrange an international workshop in Kumasi, Ghana, in September 1978, to bring together all the grantees with several internationally recognized experts and specialists in the subject of mycorrhiza. The material presented at that workshop, edited and reorganized, forms the content of this book. The main part of the book consists of five extensive reviews by international experts. Also included are summaries, of varying length, contributed by recipients of IFS grants.

The various types of mycorrhiza and host plants involved in symbiotic associations are well represented here. Most space is devoted to the ectomycorrhiza of pines, both indigenous and exotic species as well as those of indigenous species of conifers other than pines. Special attention is accorded the practical applications of mycorrhiza in forest nurseries where widespread use is most likely to be found in future.

As happens in the whole field of mycorrhiza research, endo-mycorrhiza takes second place to ectomycorrhiza. Only two IFS grants are awarded for endomycorrhiza relations in agricultural crop plants; a few others are concerned with endomycorrhiza and natural vegetation of tropical regions. This is an area where more research is needed and future IFS support for the subject should be encouraged. It is hoped that this account will provide an overview of much of the knowledge about mycorrhiza in tropical countries as well as indicating the needs for research and the application of the research findings.

The Foundation wishes to express its thanks to the advisers in the mycorrhiza programme, Professor Peitsa Mikola of the University of Helsinki, and Dr Göran Lundeberg of the Royal College of Forestry in Stockholm. Their efforts over the years in defining the programme, giving individual assistance to those receiving grants, and planning the workshop in Kumasi have been invaluable. This is also the place to record the Foundation's gratitude to the late Professor Erik

Björkman of the Royal College of Forestry, who gave the first impulse to the mycorrhiza programme.

The Foundation is deeply indebted to authorities and persons in Ghana for the arrangements at Kumasi. Particular thanks go to Dr Albert Ofosu-Asiedu for his untiring efforts in the workshop preparations. The Salén Foundation of Stockholm shared the expenses with IFS, and a further contribution from UNESCO is gratefully acknowledged. Special thanks go to Professor Peitsa Mikola for editing the present book. Its acceptance for publication by the Oxford University Press is greatly appreciated by all the contributing authors.

Sven Brohult
President of IFS

List of Contents

List of Contributors

K. Abeynayake: Department of Botany, University of Sri Lanka, Colombo, Sri Lanka.

B. K. Bakshi: Forest Research Institute, Dehra Dun, India.

S. N. Baradas: Department of Plant Pathology, University of the Philippines at Los Baños, Laguna, The Philippines.

R. Black: Department of Botany, University of Leeds, Leeds, England.

G. D. Bowen: Division of Soils, CSIRO, Glen Osmond, South Australia.

M. A. Chaudhry: Department of Forestry, Makerere University, Kampala, Uganda.

D. P. de Alwis: Department of Botany, University of Sri Lanka, Colombo, Sri Lanka.

H. G. Diem: ORSTOM/CNRS, Dakar, Sénégal.

K. Djavanshir: College of Forestry, University of Tehran, Karadj, Iran.

Y. R. Dommergues: ORSTOM/CNRS, Dakar, Sénégal.

S. A. Ekwebelam: Savanna Forest Research Station, Forestry Research Institute of Nigeria, Samaru-Zaria, Nigeria.

R. L. Ferrer: Instituto de Botánica de la Academia de Ciencias de Cuba, Habana, Cuba.

R. A. Gbadegesin: Savanna Forest Research Station, Forest Research Institute of Nigeria, Samaru-Zaria, Nigeria.

G. Germani: ORSTOM/CNRS, Dakar, Sénégal.

R. Grada-Yautentzi: Laboratorio de Microbiología Agrícola, Escuela National de Ciencias Biológicas, Instituto Politécnico Nacional, México.

S. Hadi: Department of Forest Management, Faculty of Forestry, Bogor Agricultural University, Bogor, Indonesia.

P. M. Halos: Department of Plant Pathology, University of the Philippines at Los Baños, Laguna, The Philippines.

D. S. Hayman: Soil Microbiology Department, Rothamsted Experimental Station, Harpenden, Herts., England.

R. A. Herrera: Instituto de Botánica de la Academia de Ciencias de Cuba, Habana, Cuba.

C. Iloba: Department of Crop Science, University of Nigeria, Nsukka, Nigeria.

M. H. Ivory: Unit of Tropical Silviculture, Commonwealth Forestry Institute, Oxford, England.

C. Khemnark: Department of Silviculture, Faculty of Forestry, Kasetsart University, Bangkok, Thailand.

S. Kondas: Department of Forestry, Tamil Nadu Agricultural University, Coimbatore, India.

M. Martini: Forestry Research Institute, Bogor, Indonesia.

D. H. Marx: Institute for Mycorrhizal Research and Development, USDA Forest Service, Athens, Georgia, USA.

P. Mikola: University of Helsinki, Helsinki, Finland.

M. Moawad: Institut für Tropischen und Subtropischen Pflanzenbau, Göttingen, Germany.

Z. O. Momoh: Federal Department of Forestry, Ibadan, Nigeria.

B. Mosse: Soil Microbiology Department, Rothamsted Experimental Station, Harpenden, Herts., England.

P. Nadarajah: Department of Botany, University of Malaya, Kuala Lumpur, Malaysia.

A. Ofosu-Asiedu: Forest Products Research Institute, Kumasi, Ghana.

E. Owusu-Bennoah: Department of Crop Science, University of Ghana, Accra, Ghana.

R. G. Pawsey: Forest Research Institute of Malawi, Zomba, Malawi.

J. Pichot: IRAT/GERDAT Radioagronomy Laboratory, Montpellier, France.

A. Rambelli: Laboratorio di Micologia, Istituto dell'Orto Botanico, Università di Roma, Italy.

J. F. Redhead: Division of Forestry, University of Dar es Salaam, Morogoro, Tanzania.

S. Riess: Laboratorio di Micologia, Istituto dell'Orto Botanico, Università di Roma, Italy.

N. Saleh-Rastin: College of Forestry, University of Tehran, Karadj, Iran.

M. N. Shamsuddin: Faculty of Forestry, Universiti Pertanian Malaysia, Serdang, Selangor, Malaysia.

B. Truong: IRAT/GERDAT Radioagronomy Laboratory, Montpellier, France.

M. Valdés: Laboratorio de Microbiología Agrícola, Escuela National de Ciencias Biológicas, Instituto Politécnico Nacional, México.

U. P. de S. Waidyanatha: Rubber Research Institute of Sri Lanka, Agalawatta, Sri Lanka.

A. Wild: Soil Science Department, University of Reading, Reading, England.

PART I

Introduction

1 Mycorrhizae across the frontiers

P. Mikola

Mycorrhiza, the symbiotic association of fungi and roots of vascular plants, has been known for about one hundred years. After several decades of sporadic studies and occasional heated controversy over the symbiotic or parasitic nature of the association, a great advance in mycorrhiza research was made in the 1920s, primarily through the pioneering work of the Swedish scientist Elias Melin. Melin (1923, 1925, 1936) definitely proved the symbiotic character of the ectotrophic mycorrhiza of forest trees, which, in turn, gave a strong impetus to increasing activity and even international co-operation in mycorrhiza research.

At about the same time, in the early decades of the current century, the role of mycorrhizal symbiosis was bitterly experienced in the emerging silviculture of many tropical countries. Practically oriented forest research usually began with planting trials of exotic species, and these, particularly the efforts of introducing exotic pines, very often failed. The reason for the failure was suspected, and later on proved, to be the absence of suitable symbiotic fungi which, again, could be corrected by intentionally or accidentally introducing mycorrhizal infection.

The history of the invasion of mycorrhizal infection across the frontiers is like a series of detective stories (for details, see Mikola 1970). Where pines or other ectomycorrhizal trees were first introduced as potted plants, the mycorrhizal fungi travelled in their roots and no other importation was needed. This has, apparently, been rather common, since early settlers from Europe often carried tree seedlings of their home countries and planted them around their new homes. This is probably the way in which the first pines or other ectomycorrhizal trees with their mycorrhizal symbionts arrived in South Africa, Australia, New Zealand, and South America. Mycorrhizal fungi have also been carried in tree roots for shorter distances when, on the establishment of a new nursery, transplants have been brought in from older nurseries.

Mycorrhizal infection may also spread as spores through the air or shorter distances in soil adherent to shoes of workers, tools, car tyres, etc. This may explain how mycorrhizae have often developed in new nurseries far from existing pine forests, without intentional inoculation.

There are numerous examples, however, where all efforts at growing pines consistently failed until mycorrhizal fungi were brought in and inoculated into nursery soil. Such cases have been recorded, for instance, from Puerto Rico (Briscoe 1959), Trinidad (Lamb 1956), Nigeria (Madu 1967), Malawi (Clements 1941), Kenya (Gibson 1963), and Zambia (Mikola 1970).

Because of incidental sources of mycorrhizal inoculation, the fungal population in different nurseries and planting areas may vary greatly. This was clearly demonstrated by an extensive study on the structure of pine mycorrhiza in tropical nurseries (Mikola 1978). Although great variation was noticed in the external appearance and anatomical structure of mycorrhizae, such as colour or thickness of the mantle, three main types could be distinguished.

1. The most common type was a brown ectotrophic mycorrhiza with a thin mantle and Hartig net (Fig. 1.1). The mantle could even be almost absent. This type approximately corresponds to the 'hazelnut brown form' of Rambelli (1967). It was the dominant type in East African nurseries, for instance.

2. Ectotrophic mycorrhiza with a thick (20–40 μm) and compact mantle and well-developed Hartig net occurred in Trinidad (Fig. 1.2), where the mycorrhizal inoculum originates from the natural range of

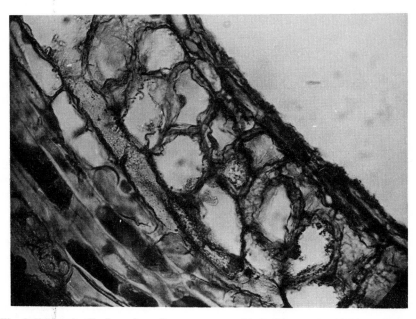

Fig. 1.1. Longitudinal section of an ectomycorrhiza of 3-month-old *Pinus patula*. The mantle is almost lacking but the Hartig net is well developed. Rwabaranda nursery, Uganda.

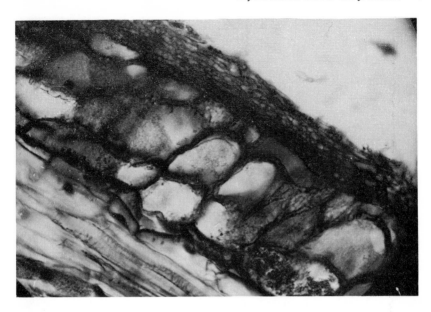

Fig. 1.2. Longitudinal section of an ectomycorrhiza of 6-month-old *Pinus caribaea*. The thick mantle and Hartig net are clearly visible. Cumuto nursery, Trinidad.

Pinus caribaea in British Honduras (Lamb 1956). The colour was light-brown. A somewhat similar mycorrhiza with a thinner (15–20 μm) mantle and very thin hyphae in the Hartig net was dominant in one nursery in Swaziland. Ectotrophic mycorrhizae with a loose, white mantle and rich external mycelia were also sometimes met although they were uncommon.

3. A typical ectendotrophic mycorrhiza with very coarse hyphae and Hartig net and heavy intracellular penetration, and almost without a mantle (Fig. 1.3), dominated some nurseries and was lacking in others. This type closely resembles the description of the 'chestnut-brown form' of Rambelli (1967). It was also very similar to the ectendotrophic mycorrhiza which is common on pine in Finnish forest nurseries (Mikola 1965), as well as elsewhere (Laiho 1965), and is probably caused by the same fungus. The degree of the intracellular penetration can vary considerably, whereas the very coarse Hartig net is the most characteristic anatomical feature. The ectendotrophic type is found, for instance, in Tanzania, Zambia, and Swaziland, and also in several comparative samples outside the tropics, e.g. in Australia, New Zealand, the Mediterranean countries, and Central Europe. The presence of the ectendotrophic infection in some nurseries and its absence in others was most conspicuous in this study.

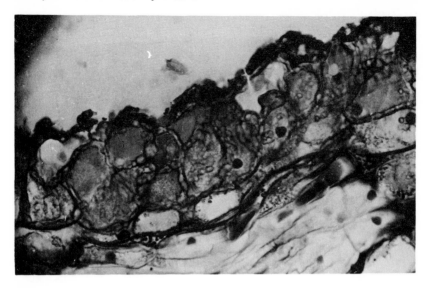

Fig. 1.3. Longitudinal section of an ectendomycorrhiza of 1-year-old *Pinus kesiya*. The very coarse Hartig net is most conspicuous. Chati nursery, Zambia.

Transportation of mycorrhizal infection from one country to another is technically easy in soil or roots of living seedlings. However, as quarantine regulations usually prohibit import of living plants and unsterilized soil, there have often been bureaucratic difficulties and, in fact, mycorrhizal inoculum has been smuggled illegally to some countries. Such cases, of course, have not been well documented.

The transport of soil or living seedlings for mycorrhizal inoculation is not at all a satisfactory method. Although the technique itself is easy, there are several drawbacks. Soil can be bulky for long-distance transport, and disadvantageous selection of the fungal population may take place during prolonged storage and transport. Furthermore, the fungal population of such a mixed inoculum is unknown; the fungi present may be less suitable for the prevailing conditions of the new site. The greatest danger, however, is the risk of introducing pests and diseases with the mycorrhizal infection.

To overcome the above drawbacks, inoculation with fungal spores or pure cultures has repeatedly been suggested (e.g. Bakshi 1967). Pure cultures have been commonly used in scientific experiments, but practical application has not been so successful. Besides technical difficulties in isolating, culturing, and inoculating mycorrhizal fungi, there are still several fundamental questions to be answered; for instance, which fungal species would be the best symbionts for different tree species and for different ecological conditions. For these reasons mycorrhizal inoculation with pure cultures has reached the stage of practical application in only a few cases. Scientists working with mycorrhizae quite often send pure cultures to each other in the same way as microbial cultures are commonly shipped across international boundaries for medical or industrial purposes.

Fungal spores or whole sporocarps have also been sent over long distances for mycorrhizal inoculation. Spore inoculation, however, has not been studied in detail, and therefore the success of occasional shipments of spores has remained largely unknown.

The first steps towards international co-operation on mycorrhizae were unofficially taken by practising foresters. When they experienced failures in their efforts to introduce exotic pines and heard rumours on the miraculous effect of nursery soil inoculation, they tried to get such inoculum, legally or illegally, from their colleagues in other countries. Scientists likewise have co-operated through correspondence and used personal contacts to obtain fungal cultures, when needed.

International co-operation in mycorrhiza research was first organized in 1956. At the twelfth Congress of the International Union of Forest Research Organizations (IUFRO) in Oxford, UK, a special Working Group was established to keep contacts and co-ordinate proposed joint projects in mycorrhiza research. Since then this open group has had regular meetings at the IUFRO Congresses and at least one other meeting, in Puerto Rico, in 1964. Although the group has been quite informal and no joint projects have been organized, its very existence has stimulated interest in mycorrhizae, and the members have kept in contact with each other through the exchange of publications and other correspondence. Some scientists from tropical countries also have participated in this group. At the last IUFRO Congress, in 1976, the group was reorganized to include other root symbioses and root physiology in general.

Besides the IUFRO group and its meetings, several international conferences have stimulated mycorrhiza research and co-operation in recent years. Such an international conference was held, for instance, at Weimar, German Democratic Republic, in 1960 (Rawald and Lyr 1963). The North American Conferences on Mycorrhizae have also become international. The first one, in 1969 (Hacskaylo 1971), was only attended by North American scientists, but the two others, in 1974 and 1977, have had a strong international participation.

Overseas studies also should be mentioned as a form of international co-operation. As there have been few eminent scientists in the study of mycorrhizae and special training has been available in only a few places, students from other countries have come to these places for advance training. As an example I would like particularly to recall Professor Elias Melin and his Institute of Physiological Botany at the University of Uppsala, where several overseas students, varying in background from India to the United States, spent periods of time in the 1930s to the 1950s.

Although endotrophic mycorrhiza has been known to scientists almost as long as the ectotrophic one, only recently has it attracted a wider international interest. It has been known for a long time that the endotrophic mycorrhiza is much more prevalent in nature than the ectotrophic one. The occurrence of ectomycorrhiza is restricted to a few plant families only, whereas endomycorrhizae exist in almost all the other plant families. In spite of that, endomycorrhiza seemed to have less practical application. In forestry, for instance, while mycorrhizal inoculation was necessary for the successful introduction of exotic pines to the tropics, apparently no inoculation was needed for cypress, an endotrophic conifer. Cultivation of ornamental orchids was the only special field, a very restricted one, where endotrophic mycorrhiza found practical application, whereas the commonest type, the vesicular-arbuscular or VA mycorrhiza, seemed to be such a ubiquitous phenomenon that it deserved little attention in agriculture or forestry, for instance.

Research in endotrophic mycorrhiza has proceeded more slowly than in ectotrophic mycorrhiza. The reasons for this may be many, e.g. since ectomycorrhiza seemed to have more practical applications, it was more attractive to scientists. The more uniform and less common ectomycorrhizae were also an easier subject to study than the morphologically and even physiologically more diversified endomycorrhizae. Thus, for instance, whereas many ectomycorrhizal fungi were already known more than 50 years ago, fungal partners of VA mycorrhizae have only been determined in the last 20 years. Research on endomycorrhizae has also encountered some methodological difficulties, e.g. in the isolation and pure culture of the VA endophytes.

Progress in endomycorrhiza research, however, has been particularly fast in recent years (cf. Mosse 1973). New findings have also opened up the prospects for practical application. Experiments have shown that the endotrophic association can greatly increase the phosphorus uptake of plants in phosphorus-deficient soil. This may give some possibility, for instance, to improve the phosphorus utilization by plants with proper manipulation of mycorrhizal association.

The number of scientists working with endomycorrhizae has greatly increased in recent years and international co-operation has developed. As evidence of this, the first international conference solely on endotrophic mycorrhizae at Leeds, UK, in 1974 (Sanders, Mosse, and

Tinker 1975) should be mentioned. Of course, endomycorrhizae were already included in the programmes of the three North American Conferences and the Weimar symposium in 1960, although by far the main emphasis at those conferences was on ectomycorrhizae.

As a whole, interest in mycorrhiza research has greatly increased in recent years. For instance, the number of participants in the First North American Conference on Mycorrhizae in 1969 was 56, whereas the corresponding number at the Third Conference eight years later was 229.

The above progress has mainly taken place in temperate countries. Mycorrhiza research, however, seems to have the greatest importance in the tropics. As was mentioned above, the role of ectomycorrhizal associations was first experienced at the introduction of exotic pines into tropical countries. Since the development of forestry in the tropics is largely based on the introduction of exotics, continued investigations into various aspects of mycorrhizal relations in both indigenous trees and exotic species are urgently needed. Inoculation of pines is probably the field of research where practical application of the results can most likely be expected, or rather the results are already applied in forest nursery practice. The occurrence of ectotrophic mycorrhiza in some tropical and subtropical tree families, such as Dipterocarpaceae, Caesalpiniaceae, and Myrtaceae, is a rather recent discovery and needs further investigation.

With regard to endomycorrhizae, very little or nothing is so far known about mycorrhizal relations in many tropical crop-plants. The recent findings on the effect of endotrophic association on the phosphorus uptake by plants are particularly important. Since tropical soils are very often deficient in phosphorus and, on the other hand, phosphorus fertilizers are expensive, phosphorus nutrition is one of the bottle-necks of tropical agriculture and even a slight improvement of mycorrhizal efficiency might have a great practical importance. The complex association of mycorrhizal fungi together with nitrogen-fixing bacteria in the roots of many leguminous crop-plants of the tropics, as has been discovered recently, is another promising area where mycorrhiza research can benefit tropical agriculture.

For basic information about tropical ecology, mycorrhizal relations in natural ecosystems also should be studied, e.g. in rain forests, savannahs, grasslands, and semideserts.

Being aware of the urgent need of promoting mycorrhiza research in the tropics, the International Foundation for Science has selected mycorrhiza as one of its priority areas and is now supporting about 20 research projects in developing countries, mainly in the tropics. To promote co-operation among scientists working in this field, an International Workshop on Tropical Mycorrhiza Research was arranged in Kumari, Ghana, from 28 August to 6 September 1978. Grantees of the Foundation and a few international experts were invited to this Workshop. An edited and revised version of the proceedings of the Workshop is presented in the following pages.

References

Bakshi, B. K. (1967). Mycorrhiza—its role in man-made forests. Doc. FAO World Symp. Man-made Forests, Canberra, 1967, Vol. 2, pp. 1031-42.

Briscoe, C. B. (1959). Early results of mycorrhizal inoculation of pine in Puerto Rico. *Carib. For.* 20, 73-7.

Clements, J. B. (1941). The introduction of pines into Nyasaland. *Nyasald agric. q. J.* 1, 5-15.

Gibson, I. A. S. (1963). Eine Mitteilung über die Kiefernmykorrhiza in den Wäldern Kenias. In *Mycorrhiza. Int. Mykorrhizasymp.* (eds. W. Rawald and H. Lyr), pp. 49-51. G. Fischer, Jena.

Hacskaylo, E. (ed.) (1971). *Mycorrhizae.* Proc. First N.A. Conf. Mycorrhizae. Misc. Publ. 1189, US Dept. Agr., For. Service, Washington DC.

Laiho, O. (1965). Further studies on the ectendotrophic mycorrhiza. *Acta for. fenn.* 79, 3.

Lamb, A. F. A. (1956). *Exotic forest trees in Trinidad and Tobago.* Gov. Print., Trinidad and Tobago.

Madu, M. (1967). The biology of ectotrophic mycorrhiza with reference to the growth of pines in Nigeria. *Obeche, J. Tree Club, Univ. Ibadan* 1, 9-16.

Melin, E. (1923). Experimentelle Untersuchungen über die Konstitution und Ökologie der Mykorrhizen von *Pinus sylvestris* und *Picea abies. Mykol. Unters. Ber.* 2, 73-331.

— (1925). *Untersuchungen über die Bedeutung der Baummykorrhiza.* G. Fischer, Jena.

— (1936). Methoden der experimentellen Untersuchung mykotropher Pflanzen. *Handb. biol. ArbMeth. II* 4, 1015-108.

Mikola, P. (1965). Studies on the ectendotrophic mycorrhiza of pine. *Acta for. fenn.* 79, 2.

— (1970). Mycorrhizal inoculation in afforestation. *Int. Rev. For. Res.* 3, 123-96.

— (1978). The structure of pine mycorrhiza in tropical and subtropical forest nurseries. *Tropical mycorrhiza.* Prov. Rep. No. 1, Int. Found. Sci., Stockholm, pp. 361-71.

Mosse, B. (1973). Advances in the study of vesicular-arbuscular mycorrhiza. *A. Rev. Phytopathol.* 11, 171-96.

Rambelli, A. (1967). Atlas of some mycorrhizal forms observed on *Pinus radiata* in Italy. Pubbl. Centro Sperim. Agric. Forest., Vol. IX, suppl.

Rawald, W. and Lyr, H. (eds.) (1963). *Mykorrhiza. Internationales Mykorrhiza-symposium* (25-30 April 1960, Weimar). G. Fischer, Jena.

Sanders, F. E., Mosse, B., and Tinker, P. B. (eds.) (1975). *Endomycorrhizas.* Academic Press, London.

PART II

Ectomycorrhiza in tropical forestry

2 Ectomycorrhizal fungus inoculations: a tool for improving forestation practices

D. H. Marx

Introduction

The need of many species of forest trees for ectomycorrhizal associations was initially observed when attempts to establish plantations of exotic pines routinely failed until the essential fungi were introduced (Briscoe 1959; Clements 1941; Gibson 1963; Hatch 1936; Kessell 1927; Madu 1967; van Suchtelen 1962). The need of pine and oak seedlings for ectomycorrhizae has also been convincingly demonstrated in the afforestation of former treeless areas, such as the grasslands of Russia and the Great Plains of the United States (Goss 1960; Hatch 1937; McComb 1938; Rosendahl and Wilde 1942; Shemakhanova 1962; White 1941).

The primary purpose for inoculating with symbiotic fungi in world forestry is to provide seedlings with adequate ectomycorrhizae for planting in man-made forests. Such treatment has proven essential in forestation of cutover lands and other treeless areas, introduction of exotic tree species, and reclamation of adverse sites such as mining spoils. Use of ectomycorrhizal fungi can be of major significance to artificial regeneration.

Most research on inoculation with ectomycorrhizal fungi has been based on two working premises. First, any ectomycorrhizae on roots of tree seedlings is far better than no ectomycorrhizae at all. Success in correcting these deficiencies has contributed greatly to our understanding of the importance of ectomycorrhizae to trees, especially as they relate to the establishment of exotic forests. Secondly, some species of ectomycorrhizal fungi under certain environmental conditions are more beneficial to trees than other fungal species. Much more research should be aimed at selecting, propagating, manipulating, and managing the more desirable fungal symbionts to improve tree survival and growth.

The majority of past work on inoculation with ectomycorrhizal fungi has been done in nurseries for the production of bare-root or containerized tree seedlings. Most future work with inoculations will undoubtedly continue to concentrate on seedling production. However, the inoculation of seed for direct seeding operations could become a very important alternative to planting seedlings, especially on the more remote sites or those of very rough terrain.

Trappe (1977), Mikola (1973), and others (Bowen 1965; Imshenetskii 1955; Shemakhanova 1962) have thoroughly reviewed past work on ectomycorrhizal inoculations. I, therefore, am free to discuss selected reports as they relate to specific procedures and concentrate on recent published and unpublished work on pure culture manipulation of ectomycorrhizal fungi.

Ectomycorrhizae are formed by fungi belonging to the higher Basidiomycetes (mushroom and puffball group), Ascomycetes, and zygosporic Phycomycetes of the Endogonaceae (Gerdemann and Trappe 1974; Trappe 1962, 1971). The host plants of these fungi are predominantly trees belonging to the Pinaceae, Fagaceae, Betulaceae, Salicaceae, Juglandaceae, and other families (Meyer 1973).

Many species of fungi may be involved in the ectomycorrhizal associations of a forest, a single tree species, an individual tree seedling, or even a small segment of lateral root. As many as three species of fungi have been isolated from an individual ectomycorrhiza (Zak and Marx 1964). Even as a single tree species can have numerous species of fungi capable of forming ectomycorrhizae on its roots, a single fungus species can also enter into mycorrhizal association with numerous tree species. Some fungi are apparently host specific; others have broad host ranges and form mycorrhizae with members of numerous tree genera in diverse families (Trappe 1962).

The majority of reports on inoculations with ectomycorrhizal fungi involve Basidiomycetes on pines, oaks, and eucalypts. Several types of natural and laboratory produced inocula and methods of application have been used through the years. Many of these procedures have proved successful, others have not. Frequently, conflicting results are encountered.

Natural dissemination of spores

No ectomycorrhizal fungus in a natural environment has been shown to complete its life cycle in the absence of mycorrhizal association (Hacskaylo 1971). There also is no evidence to suggest that these fungi grow saprophytically in a natural forest soil (Shemakhanova 1962). However, once root infection has taken place extramatrical growth of mycelium from roots into large soil volumes is not unusual. Sporophores may occur dozens of metres from the above-ground parts of the tree host.

Most ectomycorrhizal fungi produce sporophores containing numerous spores. These spores can be disseminated great distances by wind, rain, insects, small animals, etc. The greater the density and closeness of the ectomycorrhizal tree hosts to the seedling producing areas, the greater the chances are for rapid natural ectomycorrhizal

development on the seedlings. In the southern United States, ectomycorrhizae appear in the spring on nursery-grown pine seedlings as
early as six to eight weeks after seed germination. This occurs even in
nursery soil fumigated just a few days prior to seeding because these
nurseries are usually surrounded by dense stands of pine and oak
which support abundant sporophores of mycorrhizal fungi. However,
dry or cold weather often influences the fruiting habits of these fungi
causing erratic spore production and dissemination. As mentioned by
Trappe (1977), although heavy rains during the fruiting season stimulate fruiting, these rains can also restrict dispersal by washing spores
from the air near their source.

The cultural procedures used to produce seedlings in bare-root or
container nurseries create environmental conditions which select
certain mycorrhizal fungi adapted to these conditions. In the southern
United States, as well as other parts of the world, *Thelephora terrestris*
appears to dominate the roots of most pines and oaks grown in nursery soil (Marx, Bryan, and Grand 1970; Mikola 1970; Weir 1921)
and in containers (Marx and Barnett 1974).

In the southern United States, spore dissemination begins shortly
after the nursery soil has been fumigated and seeded (usually from
March to May) when *T. terrestris* produces sporophores and abundant
spores in adjacent forests. These spores are carried by the wind to the
fumigated soil, leach through the soil a few centimetres, and rapidly
colonize the seedling roots.

This early colonization of seedling roots by *T. terrestris* can pre-
clude colonization by other fungi which produce spores later in the
year. These other fungi may form some mycorrhizae on the seedling
roots later in the growing season, but only rarely do they dominate
the roots. To maintain superiority on the roots, a fungus that appears
first on seedlings must form mycorrhizae on short roots as rapidly as
the short roots are produced. If it fails to do so other available fungi
will infect these new roots. *Thelephora terristris* and the other fungi
which naturally occur and dominate seedling roots in nurseries have
the capacity to spread rapidly under nursery conditions.

In nurseries in the United States and many other parts of the
world, production of tree seedlings (particularly pine, oak, spruce,
fir, and eucalyptus) is not seriously affected by deficiencies of ectomycorrhizae. In 1975, for example, there were over 1.2×10^9 coniferous tree seedlings produced in 190 nurseries in the United States.
It is highly unlikely that these seedlings would have reached plantable size if they had not had adequate mycorrhizae. Ectomycorrhizal
fungus deficiencies in properly managed tree nurseries in the United
States, therefore, are rare.

The few mycorrhizal deficiencies reported in the United States in recent times (Marx, Morris, and Mexal 1978; Trappe and Strand 1969) have appeared in newly established nurseries in which the soil was fumigated to control weeds and pathogens. Fumigation also eliminates residual inoculum of ectomycorrhizal fungi.

Not all deficiencies or the erratic occurrence of ectomycorrhizae on seedling roots are due to the absence of inoculum in soil. Excessive use of soluble fertilizers can reduce susceptibility of roots to infection (Bowen 1973; Marx, Hatch, and Mendicino 1977). In addition to fumigants, certain fungicides can significantly reduce or eradicate inoculum in soil and cause a deficiency of mycorrhizae on seedling roots (Hacskaylo and Palmer 1957; Iyer, Lipas, and Chesters 1971; Wojahn and Iyer 1976). However, it should be pointed out that certain fungicides can also stimulate development of ectomycorrhizae (Marx and Bryan 1969a; Powell, Hendrix, and Marx 1968), as can the application of certain nutrients to the soil which either stimulate residual inocula or host-root suscpetibility (Bowen 1973).

Natural inoculation of containerized seedlings grown in sterile or near-sterile potting mix is undoubtedly from airborne spores, but in many instances it is very erratic (Trappe 1977). In the southern United States pines are grown in a variety of containers for approximately three to four months (Balmer 1974). Natural ectomycorrhizal development on these seedlings is often erratic because they are watered and fertilized heavily to obtain the fastest possible growth in the shortest possible time. Unfortunately, these conditions induce high shoot–root ratios and low incidences of mycorrhizae on pine, both of which are thought to be undesirable for best field performance of seedlings (Marx and Barnett 1974; Marx, Hatch, and Mendicino 1977; Ruehle and Marx 1977). To ensure good ectomycorrhizae on containerized seedlings, procedures for inoculation with specific fungi and providing nutrients and water must be perfected (Ruehle and Marx 1977; Trappe 1977).

Cultural practices in nurseries influence the incidence of specific ectomycorrhizal fungi on tree seedlings. Levisohn (1965) reported that for many years *Suillus bovinus* was the only mycorrhizal fungus on seedlings of *Picea sitchensis* in certain nurseries in England. These nurseries were not fumigated and the soils were heavily composted each year with a variety of organic residues. The *S. bovinus* mycorrhizae soon disappeared from roots of *P. sitchensis* after outplanting unless the outplanting site was also heavily composted with organic matter. These results suggest that this fungus is not adapted to soils with low levels of organic matter. During subsequent years the nursery expanded and began to use different methods of fertilization in addition to amending the soil with organic residues. Associated with those cultural changes in the nurseries was the appearance of *Rhizo-*

pogon luteolus. Eventually this fungus dominated the spruce seedlings in the nursery, and thereafter *S. bovinus* occurred only rarely.

Soil or humus containing natural inoculum

The easiest and simplest method for eliminating ectomycorrhizal fungus deficiencies on seedlings in nurseries is to apply soil, humus, or duff containing mycorrhizae and associated mycelium. It is by far the most commonly used method to ensure consistent development of mycorrhizae (Mikola 1973). This form of inoculum can be collected from natural forests, plantations, or established tree nurseries. It is a very reliable method if done properly. The use of soil inoculum to propagate and maintain specific mycorrhizal fungi is unusual, but soil from truffle producing areas was used to inoculate seedlings in new areas (Malencon 1938). The success of this effort is unknown. In the Soviet Union (Shemakhanova 1962), soil containing mycorrhizae is routinely placed in holes prior to planting acorns for regeneration of oaks in shelterbelts. These areas of the Soviet Union are not devoid of ectomycorrhizal fungi, but apparently soil inoculation increases survival and early growth of the oak seedlings. The purpose of inoculation is to introduce new fungi into the area and enhance development of mycorrhizae on seedlings. The latter point is considered critical for artificial regeneration of oaks in Russia.

Mikola (1970) discusses in detail the various practices used in tropical and subtropical countries to ensure mycorrhizal development on nursery seedlings raised in various containers (pots, wooden boxes, plastic tubes, Swaziland beds, etc.). In most instances, 10 to 20 per cent of the container mixture is topsoil or humus containing mycorrhizae collected from a healthy pine plantation or established nursery. In order to conserve soil inoculum a small quantity of soil is sometimes added to the base of individual seedlings in containers.

A major drawback to the use of soil or humus as inoculum is that the specific fungi in the mixture cannot be controlled. Also there is no assurance that the chosen soil inoculum contains the most desirable fungi for the tree species being produced. The large volumes of soil needed to inoculate a nursery creates a logistics problem since 10 per cent of the volume of soil is currently recommended to assure adequate inoculation of nurseries (Mikola 1973). This volume of mycorrhizal soil can be extremely large in a nursery covering many hectares. Soil inoculum may contain a variety of harmful micro-organisms and noxious weeds in addition to the ectomycorrhizal fungi. Some of these micro-organisms may not be potentially harmful only to the tree seedling crop (Mikola 1973) but possibly to nearby agricultural crops (Marx 1975).

Ectomycorrhizal seedlings or excised ectomycorrhizae

Tree seedlings with ectomycorrhizae or excised mycorrhizae have been used as inoculum for new seedling crops. Chevalier and Grente (1973) were able to successfully establish the truffle fungus *Tuber melanosporum* in nursery beds from seedlings previously inoculated with this fungus. New seedlings growing adjacent to the pre-inoculated seedlings formed *T. melanosporum* ectomycorrhizae. Mikola (1973) discussed the Indonesian technique of inoculation. Seedlings with abundant mycorrhizae are planted at one to two metre intervals in new seedbeds. This technique is highly successful in forming ectomycorrhizae on seedlings of *Pinus merkusii.* Levisohn (1956) used surface-sterilized pine roots with *Rhizopogon luteolus* mycorrhizae to successfully form mycorrhizae on and stimulate growth of *Pinus contorta* seedlings in pots. Ekwebelam (1973) inoculated *Pinus caribaea* var. *hondurensis* and *P. kesiya* by growing them for three months in polyethylene bags filled with sterile sand containing excised mycorrhizae formed by *Rhizopogon luteolus.* The typical white coralloid ectomycorrhizae usually associated with *R. luteolus* was observed on roots of the seedlings within one month of germination. Ekwebelam did not mention non-inoculated control seedlings in his experiment, but in his previous experiments non-inoculated seedlings grown under similar conditions in this area of Africa apparently remained free of ectomycorrhizae.

Procedures using seedlings with ectomycorrhizae or excised mycorrhizae may be useful in propagating a specific fungus if soil conditions maintained in the nursery favour the introduced fungi and not those that occur naturally. Unless the introduced fugus is more adapted to nursery conditions than the naturally occurring ones, the desired fungus will eventually be displaced from roots.

Sporophores and spores

According to Trappe (1977), the first attempts to use specific fungi to form mycorrhizae on seedlings dates back to the eighteenth century. Sporophores of truffle fungi were added to planting holes of oak seedlings in new plantations in attempts to enhance truffle production (Malencon 1938). These inoculations took place nearly 75 years before the term 'mycorrhiza' was coined and over 100 years before the true nature of ectomycorrhizal associations was demonstrated. Unfortunately there is no way of determining to what degree these inoculations were successful. Sporophores of various ectomycorrhizal fungi, such as *Pisolithus tinctorius* and *Rhizopogon luteolus,* have been dried and/or chopped into small pieces and used to infest soil successfully (Donald 1975; Fontana and Bonfante 1971; Mullette 1976). Fresh sporophores have also been added to soil inoculum prior to its use in nurseries to enhance the infective capacity of the soil (Mikola 1973). Inoculum composed of whole or chopped sporophores is basically nothing more than spore inoculum, since the vegetative matrix of the sporophores undoubtedly decomposes shortly after incorporation into soil.

Portugal

In 1970, Azvedo (personal communication) began developing a technique of seed inoculation using dried sporophores of different ectomycorrhizal fungi. Sporophores of *Amanita muscaria, A. phalloides, Boletus granulatus, B. scaber, Hydnellum zonatum, Lactarius delicious, L. chryzoreus, Lepiota procera, Russula cyanoxantha, Sarcodon imbricatum,* and *Tricholoma terreum* were collected fresh and dried carefully in the laboratory for one week. They were then transferred to a dessicator maintained at 30 °C for a few more days to complete the dehydration, crushed into a fine powder, and stored in sealed polyethylene bags. Seeds of *Pinus pinaster* were moistened with water and coated with the dried inoculum. After six to eight months in greenhouse tests using steamed soil in pots, *A. muscaria, R. cyanoxantha, S. granulatus,* and *H. zonatum* were found to be the most efficient fungi in forming typical ecto- and ectendomycorrhizae on *P. pinaster.* In another test, *R. cyanoxantha, T. terreum,* and *B. granulatus* formed the most ecto- or ectendomycorrhizae on *P. pinaster* after three months in Japanese paper pots. Control seedlings from non-inoculated seed, in most instances, remained free of any type of mycorrhiza. Azevedo states that this dried form of inoculum remains viable for four to five years when properly stored. Again, we can assume that the functional portion of the dried sporophores is basidiospores and not mycelium. These dried basidiospores of various fungi survived considerably longer than basidiospores of *Rhizopogon luteolus* in other experiments.

Australia

Pryor (1956) added basidiospores of *Scleroderma flavidum* to heat-sterilized soil in pots. Abundant ectomycorrhizae formed on roots and growth of *Eucalyptus dives, E. pauciflora,* and *E. macrorrhyncha* was stimulated. From these results he concluded that the absence of ectomycorrhizae on these *Eucalyptus* spp accounted for regeneration failures in Iraq and other parts of the world.

Theodorou (1971) found that inoculation of seeds of *Pinus radiata* with freshly harvested basidiospores of *R. luteolus* was an easy and effective way of introducing mycorrhizal fungi into both sterile and non-sterile soil (mycorrhizal fungus deficient) in pots and in the field. This technique involved soaking surface-sterilized seeds of *P. radiata* in a sterile water suspension of basidiospores which coated each seed with approximately 1.9×10^6 spores. Theodorou found that more mycorrhizae formed on seedlings grown in sterilized soil than on those grown in non-sterile soil. He concluded that sterilization of soil enhanced mycorrhizal development by eliminating soil organisms deleterious to *R. luteolus*. Theodorou and Bowen (1973) later found that spores from freeze-dried sporophores of *R. luteolus* could be used to inoculate seed. Seed coated with basidiospores could be dried and stored (2 °C) for one month. They found that spore numbers must be increased by up to 100 times with freeze-dried spores and up to ten times with spores air-dried for two days to obtain ectomycorrhizal development on *P. radiata* seedlings equal to that of freshly collected basidiospores. Apparently freeze- and air-drying kills or inhibits germination of a substantial number of these basidiospores.

In pot studies Lamb and Richards (1974a,b) found that chlamydospores of three unidentified fungi were generally not as effective as basidiospores of *Rhizopogon luteolus, Suillus granulatus,* or *Pisolithus tinctorius* in forming ectomycorrhizae on *Pinus radiata* in natural soils lacking ectomycorrhizal fungi. The effectiveness of the different types of spore inocula, however, was improved by increasing inoculum density of the fungi or by increasing the amount of available phosphorus in the soils to 40 kg/ha. This stimulating effect of phosphorus is somewhat surprising since Mullette (1976) reported that basidiospores, i.e. crushed sporophores of *P. tinctorius*, would not form mycorrhizae on *Eucalyptus gummifera* in sterile quartz containing more than 3 kg of available P/ha (5 p.p.m.).

South Africa

Donald (1975) added air-dried and ground sporophores of *Rhizopogon luteolus* to fumigated soil in South Africa prior to seeding *Pinus radiata*. After eight months, seedlings from inoculated soil in one nursery had abundant white ectomycorrhizae with loose mycelium radiating from them. Sporophores of *R. luteolus* occurred in the inoculated beds and were associated with the white mycorrhizae. Donald concluded that the functional component of the dried sporophores was basidiospores and that they (4.4×10^7 spores per m² of soil surface) can be used as inoculum to form mycorrhizae on *P. radiata* in a conventional tree nursery.

United States

In recent years basidiospores of *Pisolithus tinctorius* have been used in a variety of nursery and container tests on pines in the southern United States. Marx and Bryan (1975) added freshly collected basidiospores of *P. tinctorius* to fumigated soil in nursery microplots at a rate of $1.3 \times 10^{10}/m^2$ around two-month-old seedlings of *Pinus taeda*. *Pisolithus* formed approximately half of all the ectomycorrhizae on seedlings by the end of the growing season. At the time of soil infestation these seedlings had a few ectomycorrhizae formed by naturally occurring *Thelephora terrestris*. The identity of the different mycorrhizae is discussed later. This competition with *T. terrestris* for feeder roots may have accounted for the lack of dominance of *P. tinctorius* on the seedling roots. Competition between these fungi was observed recently on container-grown seedlings of *P. taeda*. Seedlings were inoculated at two, four, six, and eight weeks after germination with basidiospores of *P. tinctorius* (Ruehle 1980). The older seedlings which already had a few *T. terrestris* mycorrhizae at inoculation formed fewer mycorrhizae with *P. tinctorius* in the same period of time than younger seedlings inoculated with *P. tinctorius* before *T. terrestris* could colonize a substantial part of their root systems.

In these and subsequent experiments carried out by Institute scientists, the degree of ectomycorrhizal development is expressed as a percentage of all the short roots infected. Normally the introduced fungus occurs in mixtures on the roots with naturally occurring fungi. Therefore, the amount of mycorrhizae formed by the introduced fungi is expressed as a part of the total percentage of mycorrhizae formed.

Basidiospores of *P. tinctorius* have also been used in conventional nurseries in the southern United States. Following effective soil fumigation in two different tree nurseries, ectomycorrhizae were formed with basidiospores on seedlings of *Pinus taeda, P. elliottii* var. *elliottii, P. virginiana, P. clausa,* and *P. strobus* after one growing season. The freshly collected spores were incorporated into the soil at a rate of $2.55 \times 10^9/m^2$ of soil surface just prior to seeding. The success of soil infestation with the basidiospores varied. On *P. clausa* in Florida, *Pisolithus* accounted for about 12 per cent of all the ectomycorrhizae, and on *P. taeda* and *P. strobus* in North Carolina it accounted for nearly 70 per cent of all the ectomycorrhizae. Naturally occurring fungi formed the remaining mycorrhizae. *Pisolithus tinctorius* failed to dominate the root systems. More basic studies on *P. tinctorius* basidiospores (Marx 1976) revealed that even in a soil environment free of competing ectomycorrhizal fungi, it takes basidiospores at least two months after seed germination to form macroscopically detectable ectomycorrhizae and four months to stimulate growth of *P. taeda* seedlings. During this period other fungi obviously can colonize roots in a natural soil. These studies also revealed that basidiospores collected from dry, insect-free sporophores can be stored in amber bottles at 5 °C for 34 months without loss of capacity to synthesize mycorrhizae (Marx 1976). Currently these spores are used for ectomycorrhizal synthesis; they have been stored for over five years under these conditions.

In the spring of 1975 basidiospores of *P. tinctorius*, as well as mycelial inocula of *P. tinctorius* and other fungi, were successfully used to correct the erratic occurrence of ectomycorrhizal fungi in a new tree nursery in south-eastern Oklahoma (Marx *et al.* 1978). Basidiospores were added to fumigated and non-fumigated soil just prior to seeding at rates of 1.19, 3.56, and 7.13 X 10^9 basidiospores per m^2 of soil surface. Seedlings of *P. taeda* formed abundant *Pisolithus* mycorrhizae in all plots after one growing season. There were, however, no well-defined differences in the amount of *Pisolithus* mycorrhizae formed in plots initially infested with different quantities of basidiospores. Basidiospores formed about 50 per cent more ectomycorrhizae on seedlings in fumigated soil than in non-fumigated soil. In fumigated soil, 70 per cent of all the mycorrhizae on seedlings were formed by *P. tinctorius*, whereas in non-fumigated soil *Pisolithus* accounted for less than half of all the mycorrhizae. Other ectomycorrhizae were formed by naturally occurring fungi. Fumigation eradicated these latter fungi and other microbial competitors, increasing the effectiveness of the *Pisolithus* basidiospores.

Another study was installed in the same nursery in the spring of 1976 to examine different practical methods of infesting soil with basidiospores of *P. tinctorius* (Marx, Mexal, and Morris 1979). Basidiospores (stored at 5 °C for eight months) were added to fumigated soil prior to seeding by (a) mixing spores in a hydromulch (wood pulp suspended in water) and broadcasting with a tractor-drawn applicator, (b) dusting spores onto the soil surface, or (c) injecting spores into soil with a tractor-mounted injector. Two other treatments were (d) dusting spores or (e) drenching spores onto seedlings six weeks after seeding. The rate of basidiospore application in all treatments was 5.5 X 10^8 per m^2 of soil surface. After one growing season the 350 000 seedlings of *P. taeda* were lifted and evaluated. Spores mixed with the hydromulch (a) were the most effective treatment. Three-quarters of the seedlings in this treatment had *Pisolithus* mycorrhizae and these represented over one-quarter of the total formed on the seedlings. This development resulted in a 15 per cent increase in the number of plantable seedlings and stimulated overall seedling growth (fresh weight) by 25 per cent over non-inoculated controls. The next best treatment was (b), dusting spores onto the soil at time of seeding. Only one-third of the seedlings in this treatment had *Pisolithus* mycorrhizae, and these only represented about one-tenth of all the mycorrhizae on the seedlings. There were 13 per cent more plantable seedlings in this treatment and seedling fresh weights were approximately 12 per cent greater than the controls. There are problems in using dry basidiospores. During dusting the dry spores are difficult to control because breezes carry them great distances from the intended plot. This inconsistency of soil inoculation caused erratic development of mycorrhizae. All other methods of spore inoculation were not very effective. *Thelephora terrestris* formed abundant ectomycorrhizae on all seedlings in this study. Basidiospores of this fungus came from the numerous sporophores produced under pines planted adjacent to the nursery a few years earlier to provide a natural inoculum source (Marx *et al.* 1978).

Container-grown pine seedlings have been inoculated with basidiospores of *P. tinctorius* (Marx and Barnett 1974; Ruehle and Marx 1977). Root substrates such as vermiculite, peat moss, and pine bark, used in containers in the United States are successfully inoculated with mycorrhizal fungi because these substrates normally contain few microbial competitors. In greenhouse studies, equal success is achieved in forming *Pisolithus* mycorrhizae by dusting basidiospores onto seedlings (in a wind-free area) or mixing spores directly into the root substrate prior to seeding. Another promising technique is to mix basidiospores of *P. tinctorius* in the external matrix of encapsulated pine seed. For the past year the forest division of Hilleshög Seed Company Ltd., Landskrona, Sweden, and our research group in Athens, Georgia have been working co-operatively on the development of this technique. Encapsulation permits many spores to be placed on individual seed. However, the encapsulating material must be non-toxic to the spores and to the seed, and it must degrade rapidly after planting to permit satisfactory spore release onto the root zone.

It is obvious that spores of ectomycorrhizal fungi can be used in a variety of ways to either infest soil or inoculate seed for mycorrhizal development on seedlings in nurseries and containers. Results are not always positive, however. During the past eight years (Marx, unpublished data) basidiospores of a variety of fungi have been carefully collected, stored briefly, and used to infest steamed or fumigated soil in a special mycorrhizal fungus-free growth room in Athens (Marx 1973). With the exception of *P. tinctorius* and *T. terrestris*, basidiospores of *Amanita muscaria, A. caesarea, A. rubescens, Paxillus involutus, Lactarius deliciosus, L. piperatus, L. indigo, Laccaria laccata, Suillus luteus, Clitocybe nuda,* and *Russula emetica* did not form mycorrhizae on *Pinus taeda* or *P. echinata* seedlings in a four- to six-month test period. Trappe (1977) has also encountered difficulties in forming mycorrhizae on western conifers with basidiospore inoculum of various fungi collected in the Pacific Northwest of the United States, as has Shemakhanova (1962) in Russia with various tests on oak. Obviously, a great deal more research is needed on collecting, storing, handling, and inoculating procedures for spores of ectomycorrhizal fungi before they can be successfully used in inoculation programmes.

Advantages and disadvantages

There are advantages and, unfortunately, certain disadvantages in using spores of ectomycorrhizal fungi for inoculation purposes. The major advantage is that they require no extended growth phase under aseptic conditions in the laboratory as does the production of vegetative mycelial inoculum (see later discussion). Another advantage is their lack of bulk. According to Donald (1975), there are approximately 11 million spores per gram of ground sporophores of *Rhizopogon luteolus*. Marx and Bryan (1975) report approximately 1.1×10^9 spores of *P. tinctorius* per gram of basidiospores. Large numbers of basidiospores can be collected from mature sporophores of many ectomycorrhizal Gasteromycetes such as *Pisolithus*, *Rhizopogon*, and *Scleroderma*. In less than 12 man hours, we have extracted 12 kilograms of basidiospores of *P. tinctorius* from sporophores collected from under young loblolly pines growing on a kaolin spoil in central Georgia. This one collection contained 12.5×10^{12} basidiospores. If these spores were used at a rate of 5.5×10^8 spores/m^2 of soil surface, this collection could be used to inoculate 5.5 million seedlings. It would be nearly impossible to collect this quantity of spores from any of the other ectomycorrhizal fungi, especially those belonging to the Agaricales or Aphyllophorales, but rapid collection is an advantage of *Pisolithus*. Another advantage of spores, at least those of certain fungi, is that they can be stored from one season to the next. This is important since spores collected in the summer or early autumn would normally have to be stored until the following spring if they are to be used to inoculate nursery seedlings.

There are also certain disadvantages in the use of spores as inoculum. Spores of many fungal species cannot be germinated to determine their viability. Hile and Hennen (1969) reported low germination of basidiospores of *P. tinctorius* on agar plates and were unable to make successful single spore transfers to new media. Lamb and Richards (1974c) tested different conditions of pH, temperature, and relative humidity and found that under the best test conditions only 0.38 per cent of the basidiospores of *P. tinctorius* would germinate. Basidiospores of other fungi, however, germinated much better than those of *P. tinctorius*. For years (Marx, unpublished data) various physical and chemical stimuli were used to germinate basidiospores of *T. terrestris* and *P. tinctorius* without success. Apparently, synthesis of mycorrhizae is the only reliable means to determine viability of different spore collections of *P. tinctorius*. However, precise quantification of viable spores is difficult using the synthesis procedure (Marx 1976).

Other disadvantages are that the quantity of sporophores of many fungi required to inoculate nurseries may not be available every year and spore collections are frequently contaminated with various microorganisms. This is especially true of collections from Gastromycetes such as *P. tinctorius* where the basidiospores are exposed to the elements for several days or weeks during their maturation. Although data are not available, these contaminants may affect the health of tree seedlings or viability of spores.

The biggest disadvantage of using spores to inoculate seedlings is that it takes them several weeks to form mycorrhizae. This infection process is much slower than that achieved with mycelial inoculum (Marx, Bryan, and Cordell 1976; Theodorou and Bowen 1970). During this period of ingress less desirable fungi, such as pathogens (Marx 1972) or other ectomycorrhizal fungi, can colonize the roots and reduce the effectiveness of the introduced spore inocula. However, in parts of the world where the occurrence of ectomycorrhizal fungi is erratic or deficient, this delay may not have a significant effect on the final amount of mycorrhizae developed on tree seedlings from spore inoculum.

Mycelial inoculum

Ectomycorrhizal fungi as a group are difficult to grow in the laboratory. Many have never been isolated and grown in pure culture. Some species that have been isolated grow slowly, others often die after a few months in culture. Most ectomycorrhizal fungi require specific growth substances, such as thiamine, biotin, and simple carbohydrates, and are very sensitive to growth inhibiting substances (Palmer 1971).

The use of pure mycelial cultures of ectomycorrhizal fungi has been repeatedly recommended (Bowen 1965; Marx 1977a; Mikola 1973; Shemakhanova 1962; Trappe 1977) as the most biologically sound method of inoculation. Unfortunately, large scale nursery application of pure mycelial cultures has been severely hampered by the lack of sufficient amounts of inoculum. It may be possible to produce sufficient inoculum for research studies in small containers, pots, microplots, or even small nursery plots, but it is something else to produce a sufficient quantity of mycelial inoculum of an ectomycorrhizal fungus for a large nursery.

Another problem with pure mycelial cultures is knowing which fungal species to use under different conditions or with different hosts. During the past two decades a great deal of data on differences between mycorrhizal fungi and their differential effects on trees has been published (Bowen and Theodorou 1973; Marx 1977a; Theodorou and Bowen 1970). The first step in any nursery inoculation programme, therefore, must be the careful selection of suitable fungi (Bowen and Theodorou 1973; Mikola 1973; Trappe 1977).

Several researchers in various parts of the world have developed cultural procedures for producing pure mycelial inoculum of a variety of fungi for research purposes. In the last couple of decades, some of these procedures have been extensively used for various small experiments. Published information is available from Austria, Argentina, Australia, and, more recently, the United States. Experiments with pure culture inoculations have also been conducted in the Soviet Union, but details of these procedures or results have not been described (Levisohn 1958; Lobanow 1953; Mikola 1973). According to Wilde (1971), the use of pure cultures in the Socialist Republics in Europe have failed to produce significant results due to indigenous ectomycorrhizal fungi distributed throughout the soils. There are also numerous experiments with inconsistent or negative results. Since scientists tend to publish only positive results, experimental failures probably occur more frequently than we know from reviewing the literature (Mikola 1973).

Austria

Techniques used in Austria are based primarily on the work of Moser (1958a,b,c,d, 1959, 1961, 1963, 1965). Apparently techniques were developed initially to inoculate seedlings of *Pinus cembra* with low temperature strains of *Suillus plorans* in the nursery. This fungus was absent from the warmer nursery soils in the valley and in the alpine meadows, but it is a highly desirable fungal symbiont for the reforestation of this pine on the cold, high elevation sites near the timberline. Reforestation of these high elevation, mountainous areas is desirable in order to prevent avalanches and landslides.

For production of inoculum, *Suillus plorans* is first grown on Moser's (1958b) nutrient solution in small flasks for several days. The mycelium is transferred to 10-litre tanks containing the same nutrient solution and aerated for two to three hours daily for three to four months. The mycelium and liquid are poured into 5-litre flasks containing sterilized peat moss and fresh nutrient solutions. During the next few months, *S. plorans* grows throughout the substrate; the inoculum is then ready for use. Although attempts are made to maintain these cultures in aseptic condition, contaminations

by *Penicillium, Mucor,* and bacteria often occur. Moser (1963) refers
to this contaminated inoculum as 'half-pure cultures' and claims that
on certain occasions it proves more effective in forming mycorrhizae
than pure cultures. He speculates that these contaminants add a
'rhizosphere effect' to the inoculum which is beneficial to ectomycorrhizal development and seedling growth. He also found that
the most effective inoculum of *S. plorans* and other fungi is produced
in organic materials such as sterile forest litter, ground peat, or sawdust. He observed best results with ground peat. Very inconsistent
results were observed with agar inoculum or mycelial suspensions.
Other workers (Ekwebelam 1973; Levisohn 1956; Mikola 1973;
Marx, unpublished data) have used agar inoculum or mycelial suspensions with varying degrees of success.

Inoculum removed from the culture tanks is packaged in sterile
polyethylene bags, transported to the nursery, and when possible
applied to nursery soil within three days. The inoculum is usually
placed in 10 cm deep furrows in the soil at a rate of 3 to 4 litres of
inoculum per m^2 of soil surface. Young (one-month-old) seedlings of
P. cembra are then transplanted into these furrows. The best mycorrhizal development occurs on seedlings growing in soil previously
sterilized with heat or formalin.

Moser (1959) reported other means of using pure mycelial inoculum. It can be broadcast 1 cm deep onto soil and then chopped 10 cm
deep into the soil prior to seeding. This method requires much larger
amounts of inoculum (8–10 l/m^2 of soil surface) and the inoculum
often dehydrates on the soil surface prior to its incorporation. With
this method the inoculum must be able to survive in soil for the
elapsed time between seeding and when short roots are formed on
seedlings. With certain tree species in Austria, ectomycorrhizae may
not form for eight months following soil infestation and seeding.
This means that inoculum added to soil at sowing must be able to
survive a rather long period in the absence of a host. Transplanting
seedlings into furrows containing inoculum is preferred because it
eliminates this problem. Since transplanted seedlings of certain tree
species must remain in the nursery for two years (*P. cembra* is grown
in the nursery for up to four years), there is the option of only
inoculating every third or fourth row of seedlings. According to
Moser (1963), once root infection has taken place the introduced
fungi spread to adjacent seedlings. Another method of soil inoculation is placing inoculum in furrows between rows of established seedlings. This method is successful, but not recommended because
digging furrows near seedlings can damage roots.

Moser (1963) has also used this technique to produce mycelial
inoculum of *Suillus placidus, S. grevillei, S. aeruginascens, Paxillus*

involutus, Amanita muscaria, and *Lactarius porninsis*, either alone or
in mixtures. Although Moser only presented limited quantitative data
from different fungi/tree species tests in nurseries (1958a,b,c,d, 1959,
1961, 1963, 1965), the results show the biological significance of the
inoculation. In one of Moser's nursery tests in a sandy alluvial soil,
pure mycelial inoculum of *Phlegmacium glaucopus* formed abundant
ectomycorrhizae on spruce. The inoculated seedlings had a healthy
green colour and were considerably larger than non-inoculated seedlings, which had chlorotic foliage and roots completely free of ectomycorrhizae. Moser failed to mention the form of inoculum used,
the method of inoculation, or the duration of the nursery test. In
other tests (Moser 1963), larch seedlings were grown in both sterilized and non-sterilized soil of different types inoculated with *S.
grevillei, S. aeruginascens, L. porninsis,* and a mixture of the three
fungi. A fifth treatment was a control without inoculation. After two
years the non-inoculated seedlings in the non-sterile soil from a
spruce forest had an ectomycorrhizal frequency of 56, which was
similar to that of seedlings from non-sterile soil inoculated with the
fungi. The mycorrhizal frequency of the fungal treatments varied
from 53 to 72. However, the non-inoculated seedlings in sterilized
soil did not form ectomycorrhizae, while those in the fungal treatments had mycorrhizal frequency rates of 18 to 79. These data
proved the importance of soil sterilization as a prerequisite to the
effective use of mycelial cultures. Although no mention was made of
the type of soil sterilization used, it obviously was successful in eliminating the indigenous symbiotic fungi. In a similar test using soil
from a meadow, Moser (1963) found a much lower mycorrhizal
frequency on inoculated seedlings from the same fungal treatments,
especially on those growing in non-sterilized soil. The meadow soil
not only had fewer indigenous symbiotic fungi, but also had a reduced
potential for mycorrhizal development following inoculations with
the fungi. Moser (1959, 1963) discussed other nursery experiments
but did not provide data on seedling growth or mycorrhizal development.

Recently in Austria, Göbl (1975), a co-worker of Moser, discussed
the selection and culture of specific ectomycorrhizal fungi for nursery inoculations and methods of producing sufficient amounts of
inoculum for practical use. She generally follows the procedures of
Moser and recommends growing the fungi in a liquid medium until
adequate mycelium is obtained. This mycelium is placed in 1-litre
bottles containing cooked and sterilized cereal grains such as wheat or
white millet. Calcium sulphate (0.4 to 0.5 g/100 g of grain) is added
to improve the growth of certain fungi. These grain cultures are
shaken lightly each week and after two to four weeks at 20 to 22 °C

the grains become thoroughly colonized by the fungi. The grain cultures can then be stored at 4 to 6 °C for up to nine months. Göbl recommends that the grain culture be checked periodically for microbial purity on an appropriate agar medium.

The grain cultures are added to enriched peat moss for the final stage in the production of inoculum. The peat must be enriched with nitrogen and carbohydrates (ammonium tartrate, asparagine, soyabean meal, blood meal, malt extract, glucose), as well as inorganic nutrients in different combinations. The kinds and amounts of these supplements vary according to the species of fungus grown. Usually 7 to 10 grams of glucose per litre of peat is used as a standard for carbohydrates. Ten to 15 litres of sterile, enriched peat moss is placed in large transparent plastic bags and inoculated with a generous supply of grain culture. The plastic bags are plugged with cotton to provide aeration and are shaken occasionally during storage at 20 to 22 °C. After three to six weeks the inoculum is ready for use in the nursery. Contaminated cultures are apparently discarded. This method has been used to produce inoculum of *Suillus plorans, S. grevillei, Boletinus cavipes, Amanita muscaria,* and *Hebeloma crustuliniforme.* An interesting idea presented by Göbl (1975) was that the last phase involving the sterile peat moss could be done at the nursery. The problems created by shipping large volumes of peat moss inoculum would be eliminated by shipping just the grain cultures to the nursery.

After satisfactory inoculum has been produced it can be used to infest soil according to the various procedures of Moser (1963). Göbl (1975) recommended another unique method, which is to transplant a young tree seedling directly into inoculum contained in a larger volume of peat moss. After the peat moss supporting the seedling becomes colonized by the introduced fungus it is used for inoculum. Göbl (1975) prefers this form of inoculum to forest litter because it eliminates the introduction of unknown microbial populations into the nursery.

Argentina

Techniques used in Argentina were developed by Takacs (1961, 1964, 1967) at the Mycorrhiza Laboratory of the Instituto Nacional de Tecnologia Agropecuaria (INTA) at Castelar. When new pine nurseries are established in formerly treeless areas lacking ectomycorrhizal fungi the soil is inoculated with pure mycelial cultures. Techniques are very similar to those developed by Moser in Austria. Basidiospores or pieces of tissue from the sporophores are cultured on an appropriate agar medium. The mycelium is transferred to liquid culture, incubated, and added either to sterilized, germinated grains of cereals (such as barley), the cereal chaff, a mixture of grain and chaff, or sterilized peat moss. All substrates are enriched with a liquid medium. Takacs (1967) inoculated the substrates either with mycelial agar discs or mycelium from liquid culture. This inoculum, regardless of the physical media, is used after one to two months' incubation at room temperature. Peat moss is used more commonly than the other substrates. Pure mycelial cultures of *Amanita verna, Suillus granulatus, S. luteus, Hebeloma crustuliniforme,* a *Russula* sp, *Scleroderma verrucosum,* and *S. vulgare* have been produced by this method and are apparently available from the INTA in Argentina (1967). The details for large scale nursery inoculation in Argentina using Takacs's method were described by Mikola (1969). Usually five 200-ml flasks of each of four different fungi contained in either peat moss or grain-chaff inoculum are sent from INTA to a nursery. Upon arrival at the nursery the contents of each of these 20 flasks are mixed with 4 to 10 kg of sterilized soil or forest litter. These mixtures are kept moist and incubated for three weeks before use in the nursery beds. Using this method, twenty 200 ml 'starter' cultures can be used to produce 100 to 200 kg of soil inoculum. According to Mikola (1969), this is sufficient to infest 500 m^2 of nursery soil. Inoculum is usually added to the soil during preparation of the nursery beds.

Since quantitative data are not available for this work it is difficult to evaluate the success of this method on a large nursery scale. However, based on our current knowledge it is difficult to believe that these fungi can grow saprophytically in the sterile soil or litter at the nursery. Mycelial growth must occur in the presence of competitive micro-organisms and in the absence of essential nutrients. If the starter inoculum survives the incubation in soil or litter at the nursery, perhaps all that is really accomplished is a dilution of the original inoculum. This diluted inoculum must be sufficient to effectively colonize seedling roots in these nurseries containing few, if any, ectomycorrhizal fungi which can compete with the introduced inoculum.

There is only limited quantitative data available to this author on the earlier experimental work in Argentina preceding the broad application of mycelial inoculations. In one test inoculation was done with mycelial discs from Petri dish culture and not with inoculum prepared by any of the previously mentioned methods. Takacs (1964) isolated *Scleroderma vulgare* from sporophores collected from a plantation of *Pinus taeda* and grew it on an agar medium. A nursery soil was mixed 4:1 with sand, sterilized with methyl bromide and 10 per cent formalin, and planted with seed of *P. taeda.* Sixty days after seed germination half the seedlings were inoculated with pieces of agar containing the fungus. These mycelial pieces were placed 10 cm apart and 3 to 4 cm deep in the soil. When all the seedlings were lifted and measured after ten months, the results proved the value of inoculation. Inoculated seedlings formed abundant ectomycorrhizae and the soil from which they were lifted was a grey colour, apparently caused by the grey mycelium of *S. vulgare* colonizing the soil. Total fresh weights of the inoculated seedlings were 83 per cent greater than the non-inoculated controls. The differences were highly significant based on statistical analysis. Although it was not mentioned, it is assumed that seedlings in non-inoculated soil were free of mycorrhizae.

In an earlier test, Takacs (1961) used grain cultures of different fungi. Pure cultures of *Suillus granulatus, Scleroderma vulgare, Amanita phalloides,* and a *Russula* sp were obtained by germinating the spores on a liquid medium. The mycelium was used to inoculate sterilized, germinated barley seed. After only five days of incubation the grain cultures, according to Takacs, were ready to be used as inoculum. *Pinus pinaster, P. radiata,* and *P. thunbergii* were seeded in non-sterilized soil in a nursery bed. After 30 days, one or two grains of barley colonized by the specific fungi were placed 20 cm apart in the row next to the seedlings roots. Grain cultures of the four fungi were placed alternately in the rows. Apparently the seedlings were harvested several months later in the autumn. No quantitative data on mycorrhizal development was presented in this report but numerous ectomycorrhizae were illustrated. Takacs (1961) simply stated that the inoculated seedlings had exceptional growth and were considerably larger than the non-inoculated seedlings. No mention was made of other ectomycorrhizal fungi. It is difficult for this author to understand how these fungi spread rapidly enough from grain cultures placed so far apart to produce mycorrhiza on all seedlings. Another test was installed the following spring in different nurseries using the same techniques. Grain cultures of these four fungi were mixed together into non-sterile soil of three nurseries just prior to seeding. Germination began normally, but in two nurseries damping-off destroyed nearly 50 per cent of the seedlings in inoculated plots. The grain cultures probably contributed to the development of the damping-off micro-organisms. The seedlings were evaluated after six months. Takacs stated that the larger green seedlings had abundant ectomycorrhizae and the smaller chlorotic seedlings lacked mycorrhizae. However, he failed to report whether the large seedlings with ectomycorrhizae came from inoculated plots.

Australia

Theodorou (1967) developed pure mycelial inoculum of *Rhizopogon luteolus* using techniques similar to those of Moser. The purpose of inoculation with *R. luteolus* was to correct the deficiency of ectomycorrhizal fungi in some Australian soils and also to produce seedlings from *Pinus radiata* with a root system having a greater capacity to absorb phosphorus from soil. Earlier work by Bowen (1962) showed that *P. radiata* seedlings had a better uptake of phosphorus with mycorrhizae formed by *R. luteolus* than with other fungi. Pure cultures were produced in a medium of vermiculite, chaff, and corn meal in a ratio of 10:2:1 moistened with a liquid medium. The fungus was placed in bottles containing about 80 grams of medium and incubated for one month at 25 °C. Twenty-five grams of inoculum were buried 8 cm deep in soil contained in pots which had either been steamed, fumigated with different rates of methyl bromide, or non-sterilized. Certain pots were reinoculated with 10 grams of non-sterile soil. All pots were seeded with *P. radiata* and seedlings were evaluated after nine months. Since Theodorou (1967) did not use non-inoculated, sterilized soil as a control, we must assume that all ecto-mycorrhizae were formed by the introduced fungi and not from naturally occurring fungi. In steamed or methyl bromide sterilized soil that was not reinoculated with non-sterile soil, mycorrhizal development varied from 33 to 41 per cent. Ectomycorrhizal development varied from 17 to 31 per cent in sterilized soil containing *R. luteolus* as well as the non-sterile soil. Substantial increases in growth of *P. radiata* seedlings were correlated with mycorrhizal development. Best growth occurred in sterilized soil containing only inoculum of *R. luteolus*. Theodorou concluded that the effectiveness of *R. luteolus* mycelial inoculum is suppressed by antagonistic soil organisms and, therefore, recommends sterilization of soil prior to artificial inoculation with this fungus. In another greenhouse study, Theodorou and Bowen grew freshly collected cultures of *Suillus granulatus, S. luteus, Cenococcum graniforme,* and *Rhizopogon luteolus* in vermiculite medium for three weeks as described above. The inoculum was mixed into the upper 8 cm of steamed soil contained in pots. A non-inoculated, sterile soil control was used. All pots were seeded with *P. radiata.* The study was terminated after 14 months and seedlings evaluated. Ectomycorrhizal assessments were done both macro- and micro-scopically. The degree of mycorrhizal development on seedlings inoculated with *R. luteolus, S. granulatus, S. luteus, C. graniforme,* and the controls was 20, 12, 16, 2, and 6 per cent, respectively. Dry weights of seedlings with *R. luteolus* were 90 per cent greater and those with *S. granulatus* were 30 per cent greater than the other three seedling groups. *Suillus luteus* and *C. graniforme* mycorrhizae did not stimulate seedling growth. Seedlings with *Rhizopogon* mycorrhizae contained 47 to 125 per cent more phosphorus in foliage than control seedlings or those with other fungi, showing the ability of mycorrhizae formed by this fungus to enhance phosphorus absorption in *P. radiata.*

In another test, Theodorou and Bowen (1970) produced inoculum of fresh cultures of *S. granulatus, S. luteus,* and four isolates of *R. luteolus* and inoculated soil sterilized by gamma irradiation. The pots were seeded to *P. radiata* and seedlings evaluated after two years. Although quantitative data on ectomycorrhizal development was not presented, the authors stated that all inoculated seedlings had very good ectomycorrhizal development and non-inoculated seedlings lacked mycorrhizae. Over 100 per cent differences in dry weights of seedlings were obtained between fungi. The growth of all inoculated seedlings was significantly better than the controls. There was as much as 85 per cent difference in dry weight of seedlings induced by different isolates of *R. luteolus.* As a group the *R. luteolus* isolates were superior to the other fungi in stimulating seedling growth.

United States

Tests to artificially introduce pure mycelial cultures of ectomycorrhizal fungi into soil were begun in the early 1930s by Hatch (1936, 1937). He grew seedlings of *Pinus strobus* in non-sterile prairie soil in large pots. These pots were housed in a chamber filtered to exclude contamination from air-borne spores of mycorrhizal fungi. Three months after seeding the seedlings were small, yellow, unthrifty in appearance, and devoid of mycorrhizae. Half of the seedlings were inoculated with agar cultures of *Suillus luteus, Boletinus pictus, Lactarius deliciosus, L. indigo,* and *Cenococcum graniforme.* After five months, root evaluations revealed that *S. luteus* and *L. deliciosus* formed mycorrhizae on 30 per cent of the short roots and stimulated seedling growth. The other fungi apparently failed to form mycorrhizae. Non-inoculated seedlings remained stunted and chlorotic. Hatch proved that pure cultures of specific fungi could be used to correct the deficiency of mycorrhizae. Once the natural soil lacking mycorrhizal fungi was inoculated and mycorrhizae were formed, it supported normal growth of white pine seedlings.

Three decades passed before Hacskaylo and Vozzo (1967) initiated a series of inoculation experiments in Puerto Rico with pure mycelial cultures of various fungi. In one test (Vozzo and Hacskaylo 1971) pure mycelial inocula of *Cenococcum graniforme, Corticium bicolor, Rhizopogon roseolus,* and *Suillus cothurnatus* were used. These fungi were selected because they were proven symbionts and had distinctive hyphal colours which should aid in subsequent evaluations. Following Moser's (1963) general technique, they grew the fungi on agar and then in liquid media. The mycelium from liquid culture was used to inoculate polypropylene cups containing a 2:1 ratio of sterile peat moss and vermiculite moistened with a glucose–ammonium tartrate nutrient solution (pH 3.8). After 16 weeks of incubation the inoculum was flown from the USDA, Forest Service, Pioneering Research Unit Laboratory in Beltsville, Maryland, to Puerto Rico. In Puerto Rico, seedlings of *Pinus caribaea* were grown in a non-mycorrhizal condition for four months in a container nursery where they were watered and lightly fertilized. Seedlings were grown in 8 X 15 cm plastic bags filled with a 1:1 mixture of fumigated peat moss and vermiculite. The plastic bags were split and one-half cup of inoculum was placed against the exposed non-mycorrhizal roots of each seedling. The bag was closed and slipped into another container to hold the inoculum and root substrate intact. Ten months later the seedlings were measured and evaluated for mycorrhizal development. Certain seedlings were outplanted in the field; these results will be discussed later. Although they did not present quantitative data on mycorrhizal development, Vozzo and Hacskaylo (1971) reported that *Corticium bicolor, Rhizopogon roseolus,* and *Suillus cothurnatus* formed mycorrhizae. *Cenococcum graniforme* did not form mycorrhizae. Seedling height growth was correlated with ectomycorrhizal development, i.e. seedlings with the most ectomycorrhizae (*Corticium bicolor*) were the tallest. This study and others in Puerto Rico were complicated by the frequent occurrence of *Thelephora terrestris* sporophores and mycorrhizae throughout the nursery and on test seedlings. This fungus was introduced from the United States in pine duff inoculum in 1955 to correct the chronic deficiency of mycorrhizal fungi in Puerto Rico (Briscoe 1959).

Mycorrhizal Institute

In the south-eastern United States, formal research on the use of pure mycelial cultures began in 1966 at the USDA, Forest Service Laboratory in Athens, Georgia. In 1976, the research unit was given greater research latitude and was designated the Institute for Mycorrhizal Research and Development. One of several research goals of this multidisciplinary unit is to perfect existing techniques and to devise new ones for artificially inoculating tree seedlings with pure cultures of ectomycorrhizal fungi in bare-root and container nurseries. It is anticipated that these techniques will be valuable for not only correcting the erratic occurrence of ectomycorrhizal fungi in nurseries, but show the biological feasibility and practical value of manipulating and managing specific, highly desirable, ectomycorrhizal fungi on tree seedlings to improve the survival and growth of seedlings on routine and adverse reforestation sites.

Selecting fungi.   Initial research efforts were concentrated on *Pisolithus tinctorius, Thelephora terrestris, Cenococcum graniforme,* and a few other fungal species. *Pisolithus tinctorius* was chosen because it is readily propagated in the laboratory on a variety of agar or liquid media. It had yellow-gold hyphae and mycorrhizae which aid in its detection and quantitative assessment on seedling roots. The main reason for selecting this symbiont, however, was its apparent ecological adaptation to adverse soil conditions such as those found on coal spoils (Marx 1977a; Schramm 1966). *Pisolithus tinctorius* is also widespread on trees growing on kaolin spoils, sheet-eroded soils, borrow pits, and other biologically hostile sites. These sites are characterized by one or more adverse soil conditions—high soil temperatures, extreme acidity, high levels of Al, Mn, S, Fe, and chronic low fertility—which limit routine reforestation.

Seedlings with *Pisolithus* mycorrhizae formed in the nursery prior to outplanting on adverse sites should survive and grow better than routine nursery seedlings having *Thelephora* or other mycorrhizae. Tree seedlings 'tailored' with *P. tinctorius* should have a physiologically and ecologically adapted root system capable of surviving and persisting in adverse soils. The concept of forming mycorrhizae on seedlings with fungi ecologically adapted to the planting site parallels that proposed by Moser (1963) who used mycorrhizae formed by *Suillus plorans*, a low temperature fungus, on *Pinus cembra* to enhance reforestation of the high, cold, elevation sites in Austria.

There are results from some basic studies which help explain the persistence of *P. tinctorius* on these sites. In controlled temperature studies (Marx, Bryan, and Davey 1970) it was found that *P. tinctorius* was tolerant of high temperatures. The fungus grew in agar culture over a temperature range of 7 to 40 °C with an optimum at 28 °C. Later Momoh and Gbadegesin (1975), using a Georgia isolate of *P. tinctorius*, successfully grew mycelium of the fungus at 42 °C with an optimum at 30 °C. Lamb and Richards (1971) reported that the thermal death point for hyphae of *P. tinctorius* was 45 °C, whereas hyphae of *Rhizopogon luteolus* and other fungi were killed at 38 °C. Marx, Bryan, and Davey (1970) also reported that *P. tinctorius* formed more ectomycorrhizae on seedlings of *Pinus taeda* grown in aseptic culture at 34 °C than at lower temperatures. Later Marx and Bryan (1971) produced aseptic seedlings of *P. taeda* with different ectomycorrhizae at 25 °C and exposed them to higher temperatures. Seedlings with *Pisolithus* mycorrhiza had better survival and growth at 40 °C than non-mycorrhizal seedlings or seedlings with *Thelephora* mycorrhiza. *Pisolithus tinctorius* is also tolerant of and can be stimulated by the heavy metals often found in adverse sites such as coal spoils. Hile and Hennen (1969) found that the addition of iron and sulphur to agar medium stimulated vegetative growth of *P. tinctorius*. Muncie, Rothwell, and Kessell (1975) detected large amounts of elemental sulphur in the interior of sporophores of the fungus collected from coal spoils. These authors suggested that *P. tinctorius* has a unique, but unexplained, method of utilizing sulphur. These various studies showing the high temperature and metal tolerance of *P. tinctorius* help explain its ability to survive on coal spoils and other adverse sites.

There are several other unique features of *P. tinctorius* which make it a potentially strong fungal candidate for practical use in a variety of different forest situations. In addition to its occurrence on adverse sites, it also occurs in urban areas, orchards, and routine forest sites. It has been reported to occur in tree nurseries (Marx 1977b), especially on two- to three-year-old pine seedlings in northern nurseries of the United States (Marx, unpublished data). *Pisolithus tinctorius* has a proven tree host-range of nearly 50 species and is associated with an additional 25 tree species. These trees include most of the world's more important species. This fungus occurs in 33 countries of the world and in 38 states in the United States (Grand 1976; Marx 1977b). Problems with plant quarantine regulations regarding the use of *P. tinctorius* in most parts of the world should be minimal because of its broad, natural geographic distribution. After considering all of these features it becomes apparent that the development of inoculum and inoculation techniques with this fungus might prove useful not only in reclamation of adverse sites, but in reforestation of routine forest sites with both indigenous and exotic tree species in various parts of the world.

Thelephora terrestris was selected for testing at the Mycorrhizal Institute because it also grows rapidly in the laboratory on a variety of agar and liquid media. Its hyphae and ectomycorrhizae on pine are white to cream-brown in colour (Marx and Bryan 1970; Marx, Bryan, and Grand 1970). Unfortunately this colour characteristic is common to a number of other ectomycorrhizal fungi and, therefore, visual detection and quantitative evaluation of *Thelephora* mycorrhizae are difficult. This fungus does, however, fruit on seedling stems or adjacent to seedlings on soil in nurseries and can be readily traced from sporophores to mycorrhizae via visible hyphal strands. This fungal symbiont is also widespread on tree seedlings in nurseries, as previously described, and has a broad host range (Hacskaylo 1965; Marx and Bryan 1969b, 1970; Weir 1921). Its common occurrence in nurseries suggests that it is ecologically adapted to the good tilth, fertility, and moisture conditions of nursery soils. Obviously *T. terrestris* is one of the major ectomycorrhizal fungi on roots of the enormous numbers of tree seedlings produced in nurseries and planted out annually on the millions of hectares of reforestation and reclamation sites all over the world. Its occurrence in nurseries also indicates that it is a major competitor to introduced inoculum of other fungi for seedling roots.

Cenococcum graniforme has been tested on a more limited scale at the Institute than the previously mentioned fungi. It was selected because of its easily identified, jet-black ectomycorrhizae, broad host-range (Trappe 1964), and apparent drought and high temperature tolerance (Mexal and Reid 1973; Meyer 1964; Saleh-Rastin 1976; Worley and Hacskaylo 1959). This last characteristic suggests that it may be valuable to seedlings planted on sites having seasonal drought conditions. Unfortunately vegetative growth of *C. graniforme* in pure culture is very slow. Using pure mycelial techniques, various researchers have also encountered difficulties forming mycorrhizae with this fungus on various tree species (Hatch 1936; Marx *et al.* 1978; Theodorou and Bowen 1970; Vozzo and Hacskaylo 1971).

Pure mycelial culture. Our first attempts at producing pure mycelial cultures of *P. tinctorius, T. terrestris,* and *C. graniforme* were not very successful. None of these fungi would form mycorrhizae on seedlings of *Pinus taeda* from wheat grain cultures (Marx, unpublished data). Grain cultures were added to steamed soil in a 1:15 ratio in a mycorrhiza fungus-free growth room (Marx and Bryan 1969b; Marx 1973) and seeded to *Pinus taeda.* After five months none of the inoculated and control seedlings had mycorrhizae. Microscopic examination revealed that the grain cultures were colonized by sapro-phytic fungi and bacteria as early as three weeks after soil inocula-tion. We observed a great deal of damping-off of pine seed as did Takacs (1961). The high nutritive value of the boiled wheat contri-buted to its rapid colonization by saprophytes, which probably killed the ectomycorrhizal fungi. Our results do not support the claim of Park (1971) that grain cultures of *C. graniforme* can be used to inoculate nursery soil; our results from this and several other studies conflict with this broad recommendation. Our results with grain cultures also conflict with those of Takacs as discussed by Mikola (1973). Our grain cultures of the various fungi were decom-posed so rapidly by saprophytic organisms that we doubt if grain cultures could be used at a nursery as a starter culture to inoculate sterile soil for the production of a large volume of inoculum.

Vermiculite and peat moss moistened with a modification of Melin–Norkrans medium (Marx 1969) with glucose instead of su-crose was found to be an excellent substrate for the production of mycelial cultures of these fungi. In one of our first tests (Marx and Bryan 1970), inoculum of *P. tinctorius* and *T. terrestris* was grown aseptically for four months at 25 °C in 1 litre volumes of a 28:1 ratio of vermiculite and peat moss moistened with nutrient solution. The inoculum was mixed in a ratio of 1:8 with an autoclaved mixture of soil:peat moss:vermiculite in the growth room. Seeds of various pine species were planted, and after four months root evaluation revealed that *T. terrestris* formed mycorrhizae with 22 tree hosts and *P. tinctorius* with 14 tree hosts. All non-inoculated seedlings were free of mycorrhiza. This technique was successfully used later in the growth room in a study with *P. tinctorius* on *Pinus clausa* (Ross and Marx 1972), and again with *P. tinctorius* and *C. graniforme* on *P. echinata* (Marx 1973). In these and other studies, mycelium of *P. tinctorius* and *T. terrestris* usually completely permeated the vermiculite and peat moss particles after less than three months' incubation at 25 °C. However, because of the slow growth rate of *C. graniforme*, it is necessary to incubate this fungus for six to eight months under the same growing conditions to completely permeate the substrate.

Leaching of mycelial inoculum.   Our first tests outside the growth room using the vermiculite–peat moss inoculum in soil fumigated with methyl bromide gave variable results. After only a few weeks in the greenhouse the inoculum was extensively colonized by saprophytic fungi and bacteria in a fashion similar to that observed with the wheat grain cultures. This saprophytic colonization was markedly reduced by leaching the inoculum with water before adding it to soil. Leaching removed the non-assimilated nutrients and thus reduces the food base essential for saprophytic colonization by the micro-organisms. We compared several methods and types of liquids (physiological saline, sterile distilled water, and tap water) for leaching inoculum. The method which proved to be the best involved placing 4 litres of inoculum in a double layer of cheesecloth and irrigating this for several minutes under cool tap water. Excess water is removed by squeezing the inoculum in the cheesecloth by hand. This reduces the inoculum volume by one-third. In addition to non-assimilated nutrients, a great deal of pigment (a rich brown pigment in the case of *P. tinctorius*) and small vermiculite particles are washed from the inoculum during leaching. Other researchers (Moser 1963; Takacs 1967; Theodorou and Bowen 1970; Vozzo and Hacskaylo 1971) did not leach mycelial cultures prior to soil inoculation and certainly must have encountered competition problems with colonizing micro-organisms.

Leached vermiculite–peat moss inoculum of *P. tinctorius* was used successfully in 1972 to form mycorrhizae on pine at a microplot tree nursery (Marx and Bryan 1975). Inoculum was mixed at a ratio of 1:8 with fumigated soil contained in wooden framed microplots. Control soil was infested with autoclaved mycelial inoculum of *P. tinctorius* to standardize soil fertility and tilth. The soil was planted with seed of *P. taeda* in April and mulched. Periodic examinations of seedlings revealed that the mycelial inoculum of *P. tinctorius* was effective in forming mycorrhizae on seedlings one month following seed germination. Numerous sporophores of *P. tinctorius* were produced in these plots during August to October. After the seedlings became dormant in December they were lifted and evaluated. The mycelial inoculum formed the distinctive gold-yellow ectomycorrhizae of *Pisolithus* on 92 per cent of the feeder roots which induced more than a 100 per cent increase in total dry weights of seedlings over the controls. The control seedlings had 45 per cent of their feeder roots colonized with the cream-brown ectomycorrhizae characteristic of *Thelephora terrestris*. The visual estimate of the amount of mycorrhizae formed by each of the fungi was confirmed by reisolation of the respective fungi from the mycorrhizae and also by a fluorescent antibody technique developed for each of the fungi (Schmidt, Biesbrock, Bohlool, and Marx 1974). From these experiments it was found that the mycorrhizae formed by these two fungi could be visually assessed accurately on intact seedlings with the unaided eye. It should be pointed out, however, that only one other ectomycorrhizal type was present on seedlings. It occurred infrequently and was a pure white, coralloid type, easily distinguished from either *P. tinctorius* or *T. terrestris* mycorrhiza.

Survival of mycelial inoculum. A major concern in the use of pure mycelial cultures of ectomycorrhizal fungi expressed by other researchers (Moser 1963; Takacs 1967) is survival of inoculum in soil. Mycelial inoculum of *P. tinctorius* will survive in soil under a variety of conditions. During early months of the 1972 study (Marx and Bryan 1975), particles of the inoculum were removed periodically from the soil. Mycelium of *P. tinctorius* with good structural integrity was observed microscopically to be abundant in the laminated structure of the vermiculite particles. Mycelium within the leached vermiculite particles is apparently protected from environmental extremes and extensive saprophytic colonization. Residual inoculum will also survive in soil after overwintering. After the seedlings were removed from the microplots in December, soil which was initially infested with the mycelial cultures of *P. tinctorius* was left fallow until the following spring. Soil temperatures dropped to –7 °C on several occasions during the winter. In April soil in the plots was mixed with freshly fumigated soil at different ratios, seeded with *P. taeda*, and in December the seedlings were lifted and evaluated. Seedlings in non-diluted soil formed mycorrhizae with *P. tinctorius* on 75 per cent of the feeder roots. In the 1:1, 1:2.5, 1:5, and 1:10 dilutions, 60, 35, 25, and 15 per cent of the feeder roots, respectively, were ectomycorrhizal with this fungus. *Thelephora* mycorrhizae also occurred on these seedlings.

Another test of the persistence of vermiculite–peat moss inoculum of *P. tinctorius* was done in a pot study in the greenhouse (Marx, unpublished data). Leached inoculum was added to fumigated soil in a 1:8 ratio and incubated without a tree host at soil temperatures of 15, 20, 25, and 30 °C. After two weeks, one, two, and three months at the different temperatures, the infested soil was planted with seed of *P. taeda* and then incubated at 25 °C. Six months later the seedlings were evaluated. The seedlings in soil originally incubated for three months at 30 °C without a tree host had nearly as many ectomycorrhizae formed by *P. tinctorius* as seedlings in soil originally incubated without a host for only two weeks at 15 °C. We concluded from these studies that leached inoculum of *P. tinctorius* can survive in soil under a variety of conditions.

Nursery tests with mycelial inoculum.   Following the test in the microplots (Marx and Bryan 1975), leached vermiculite–peat moss inoculum of *P. tinctorius* was introduced into fumigated soil at state nurseries in Georgia, Florida, and North Carolina (Marx *et al.* 1976). The volume of leached inoculum used in each nursery was 2.8 l/m² of soil surface. It was broadcast onto the soil and immediately mixed thoroughly with hand tools into the upper 10 to 12 cm of soil.

In the Georgia nursery, ineffective soil fumigation apparently precluded successful colonization of *P. taeda, P. clausa,* or *P. virginiana* seedlings by the introduced inoculum. Problems with soil fumigation were evident by the very high levels of plant parasitic nematodes, phycomycetous root pathogens, and diseased pine seedlings detected at the end of the growing season. Also, the appearance of *T. terrestris* sporophores and mycorrhizae on seedlings as early as the second month after seeding suggested that the residual inoculum of this fungus from the previous year was not seriously affected by the fumigation.

Soil fumigation in the other two nurseries was effective. In the Florida nursery, seedlings of *P. taeda, P. clausa,* and *P. elliottii* var. *elliottii* had abundant *Pisolithus* mycorrhizae as early as six weeks after seed germination. Control seedlings of all pines also had a few mycorrhizae formed by naturally occurring fungi at this time. Numerous sporophores of *P. tinctorius* were produced in all plots inoculated with mycelial cultures by August or September. Considerably more sporophores were produced in *P. taeda* and *P. elliottii* plots than in *P. clausa* plots. Seedlings were lifted and evaluated eight months after study installation. Although *Pisolithus* mycorrhizae were formed in abundance, differences in seedling growth and total mycorrhizal development were not detected. Inoculated seedlings of *P. taeda* had 72 per cent mycorrhizal development of which *P. tinc-*

torius formed about one-half. Control seedlings of *P. taeda* had 66 per cent mycorrhizal development, all of which were formed by naturally occurring fungi. On inoculated seedlings of *P. elliottii* var. *elliottii*, *P. tinctorius* formed about eight-tenths of the total mycorrhizal development of 82 per cent. Control seedlings had 73 per cent development by other fungi. *Pisolithus tinctorius* formed over half of the 40 per cent mycorrhizal development on seedlings of *P. clausa*. Control seedlings of *P. clausa* only formed mycorrhizae with naturally occurring fungi on 21 per cent of the feeder roots. There was a positive relationship between the number of sporophores and the amount of mycorrhizae produced by *P. tinctorius* on the different pine species. The fewest number of sporophores were produced in plots of *P. clausa* seedlings which also had the fewest roots colonized by *P. tinctorius*.

In the North Carolina study, mycorrhizal development early in the season on seedlings in inoculated and control plots and the late summer development of sporophores of *P. tinctorius* were similar to that observed in the Florida study. However, in the North Carolina study, stimulation of seedling growth (total fresh weights) by *Pisolithus* mycorrhizae was 140 per cent on seedlings of *P. taeda* and about 100 per cent on seedlings of *P. virginiana* and *P. strobus*. Total mycorrhizal development on all pine species was also increased significantly by mycelial inoculation with *P. tinctorius*. Inoculated seedlings of *P. taeda* had a total of 64 per cent mycorrhizal development, with over nine-tenths formed by *Pisolithus*. Control seedlings had 50 per cent mycorrhizal development by naturally occurring fungi. The inoculated seedlings of *P. virginiana* had a total development of 72 per cent with two-thirds formed by *P. tinctorius*. Control seedlings had 47 per cent of their feeder roots ectomycorrhizal. Seedlings of *P. strobus* had a total development of 47 per cent with about three-quarters formed by *Pisolithus*. Control seedlings only formed 15 per cent mycorrhizae with other fungi. This study showed that following proper soil fumigation, leached vermiculite–peat moss inoculum of *P. tinctorius* can be introduced into soil of conventional tree nurseries and form abundant mycorrhizae on roots of southern pines. The field performance of these seedlings after outplanting on different routine reforestation sites will be discussed later.

Our next nursery research (Marx and Artman 1978) involved comparing the response of *P. taeda* seedlings to inoculation with leached vermiculite–peat moss cultures of *P. tinctorius* and *T. terrestris*. Studies were installed in fumigated soil in two state nurseries in Virginia, one in the coastal plain and the other in the mountains (elevation 580 m). Inoculum of each fungus was applied at a rate of 1.08 l/m² of soil surface and mixed thoroughly into the soil. Control plots received the same amount of vermiculite. This rate of inoculum was used because results from an inoculum density study, to be discussed later, indicated that it is as effective as higher rates. After seeding in April 1975, the seedlings were grown for seven months, lifted, and evaluated.

Only two morphological types of ectomycorrhizae were observed on seedlings in both nursery tests. One type, formed by *T. terrestris*, was observed almost exclusively in *T. terrestris* inoculated plots and the control plots. The other type was formed by *P. tinctorius* and it was observed only in *P. tinctorius* inoculated plots. *Pisolithus* formed nearly nine-tenths of all the mycorrhizae (75 per cent) in both the coastal plain and the mountain nursery. In both nurseries non-inoculated control seedlings had about 46 per cent mycorrhizal development. There were four to five times more sporophores of *T. terrestris* in the plots inoculated with *T. terrestris* in both nurseries than in control plots. Sporophore production by *P. tinctorius* was not recorded, but it was observed in all *Pisolithus* plots. Seedlings from *Pisolithus* and *Thelephora* inoculated plots had 57 and 31 per cent greater fresh weights in the coastal plain nursery and 40 and 20 per cent greater fresh weights in the mountain nursery, respectively, than the controls. Even though *P. tinctorius* and *T. terrestris* formed the same quantity of mycorrhizae on the seedlings, those with *Pisolithus* mycorrhizae were significantly heavier in both nurseries than those inoculated with *T. terrestris*. This indicated that *P. tinctorius*, even though ecologically adapted to soils of low fertility and other adverse conditions, is probably more efficient than *T. terrestris* in maximizing nutrient absorption from soil. It may be that mycorrhizal fungi adapted to poor soils make more efficient use of available nutrients than other fungal symbionts adapted to better soils.

In North Carolina, Krugner (1976) examined the interaction of soil fertility with these two fungi in closer detail on *P. taeda* in a microplot study. Using leached vermiculite–peat moss inoculum of each fungus, fumigated soil was infested with either *P. tinctorius*, *T. terrestris*, an equal mixture of inoculum of both fungi, or autoclaved inoculum of both fungi (control). The inoculum was standardized at 2 l/m^2 of soil surface for all treatments and mechanically mixed into the upper 10 to 12 cm of soil. Fertility treatments of N at 145 kg/ha, NPK at 145, 50, and 100 kg/ha, respectively, and no added fertilizer were imposed on the fungal and control treatments. After eight months the seedlings were lifted and evaluated. Inoculation of soil with either fungus alone or in mixture did not markedly affect seedling growth. Independent of the fungi, both fertilizer treatments significantly increased seedling growth in comparison to non-fertilized seedlings. Both fertility treatments stimulated the development of *Pisolithus* mycorrhizae whether *P. tinctorius* was added to soil alone or in mixture with *T. terrestris*. *Pisolithus* formed about one-fifth of the total mycorrhizal development of 64 per cent in non-fertilized soil, about one-half of the total development of 80 per cent in the nitrogen-treatment, and about two-thirds of the total mycorrhizal development of 80 per cent in complete NPK treatment. Seedlings in *T. terrestris* and control plots had between 55 per cent and 62 per cent mycorrhizal development in all fertility treatments, including the non-fertilized controls. Krugner concluded that *T. terrestris* did not compete well with *P. tinctorius* for seedling roots under conditions of abundant nutrient availability. He suggested that the inoculum of *T. terrestris* may not have been as vigorous as that of *P. tinctorius*. This latter point is undoubtedly a factor to consider in work of this type. However, *P. tinctorius* may simply be able to compete for roots better than *T. terrestris* in soils with good fertility, at least for the first growing season. Different results may be obtained in soil of higher fertility or in nursery studies of longer duration.

In the previous microplot and tree nursery studies natural recolonization of the fumigated soil by wind-disseminated spores of ecto-mycorrhizal fungi indigenous to the areas was very rapid and efficient. Usually within a few weeks after seed germination, mycorrhizae were formed on seedlings in previously non-inoculated soil by naturally occurring fungi. Competition existed, therefore, between the natural spore inoculum of the indigenous fungi and the inoculum of the artificially introduced fungi very early in the growing season. In 1974 we were able to examine the significance of soil inoculation with pure mycelial cultures to seedling growth and mycorrhizal development in a new tree nursery having minimal competition from native ectomycorrhizal fungi (Marx *et al.* 1978). The nursery, located in south-eastern Oklahoma, was established on former pasture land in an area surrounded by only a few scattered ectomycorrhizal trees. In 1974 the first crop of *Pinus taeda* seedlings was grown in non-fumigated soil. Recommended rates of fertilizer and pesticides were applied during the growing season. By mid-July the seedlings were stunted and chlorotic; less than 10 per cent of the seedlings had a trace of mycorrhizae. By mid-August thousands of seedlings were dying each week. More fertilizer and pesticides were added but seedling mortality continued. In January 1975, the nearly seven million seedlings were lifted and evaluated. Only 4 per cent had acceptable stem diameters (greater than 3 mm), none met the 15 cm height requirement for planting out, and very few seedlings had mycorrhizae. The nursery managers concluded that the poor growth of seedlings resulted from an insufficient quantity of mycorrhizae.

In April of 1975, a comprehensive study was installed in a new section of this nursery using pure mycelial cultures of *P. tinctorius, T. terrestris,* and *C. graniforme* in both fumigated and non-fumigated soil. Vermiculite–peat moss cultures of *P. tinctorius* and *T. terrestris* were grown for three months and the total volume of each culture vessel was leached and used as inoculum. However, the slower growing *C. graniforme* did not colonize all the mixture in three months; therefore only that part of the substrate with obvious mycelium of *C. graniforme* was leached and used as inoculum. Mycelial inoculum of all fungi was broadcast at a rate of 1.08 l/m^2 of soil surface and mixed into the upper 10 to 12 cm of soil. Control plots received the same rate of vermiculite. *P. taeda* was seeded in April, and during the growing season all seedlings received the same amount of fertilizer and water.

Approximately six weeks after seeding, ectomycorrhizae of *P. tinctorius* and *T. terrestris* were observed on seedlings in their respective plots in both fumigated and non-fumigated soil. Seedlings in other plots had only a few mycorrhizae at this time. By mid-August sporophores of both fungi were detected in their respective plots in fumigated and non-fumigated soil. Seedlings in these plots were vigorous and were nearly twice as large as seedlings in other plots. Mid-season examination of these vigorously growing seedlings showed they had 35 to 40 per cent of their feeder roots mycorrhizal with the respective fungi. Only a few black mycorrhizae of *C. graniforme* were observed on seedlings in the *C. graniforme* plots at this time. By early October, seedlings in the control non-fumigated plots and the fumigated and non-fumigated plots of *C. graniforme* began to grow at normal rates. Concurrent with this new growth was the appearance of *Thelephora* mycorrhizae on the control seedlings and *Thelephora* and *C. graniforme* mycorrhizae on the seedlings in the *C. graniforme* plots. Non-inoculated seedlings in fumigated soil were still stunted and had few mycorrhizae.

In December the 54 000 seedlings in this study were lifted and representative ones evaluated. In fumigated soil, the number of plantable seedlings (greater than 12.5 cm in height and 3 mm in root collar diameter) was increased by 155 per cent over the controls with *T. terrestris*, 140 per cent with *P. tinctorius*, and 77 per cent with *C. graniforme*. The former two fungi also increased the number of plantable seedlings in non-fumigated soil over the non-inoculated controls. Non-fumigated control plots had over twice as many plantable seedlings as fumigated control plots. None of the fungal treatments significantly increased seedling size in non-fumigated soil. However, in fumigated soil *T. terrestris* and *P. tinctorius* increased total fresh weights of plantable seedlings by 125 per cent and *C. grani-*

forme by 24 per cent. Mycorrhizal development was also affected by soil fumigation. In fumigated plots *Pisolithus* mycorrhizae accounted for eight-tenths of the total 60 per cent development. In non-fumigated soil, total development was about 55 per cent with *Pisolithus* forming over two-thirds of these. Since *Thelephora* occurred naturally in this nursery, it was difficult to make accurate assessments of the value of inoculation with this fungus. Naturally occurring fungi other than *T. terrestris* were less frequent on seedlings on fumigated soil inoculated with *Thelephora* than in non-fumigated soil, but total mycorrhizal development was significantly greater in fumigated (59 per cent) than in non-fumigated (48 per cent) soil. There was just as much naturally occurring *Cenococcum* mycorrhizae on seedlings in the non-fumigated control plots as in the fumigated soil inoculated with *Cenococcum*. *Cenococcum* mycorrhizae accounted for about one-quarter of the mycorrhizae in fumigated inoculated plots and accounted for one-eighth of the mycorrhizae in the non-fumigated inoculated plots. All of the above mycorrhizal assessments of specific fungi were confirmed by surface sterilizing the mycorrhizae and reisolating the fungi on agar medium.

The results of this study revealed several salient points. Mycorrhizal development must occur early in the growing season in order to improve the numbers of plantable seedlings of *P. taeda* and their size. Pure mycelial cultures of specific fungi can be used to correct the erratic occurrence or deficiency of mycorrhizae in both fumigated and non-fumigated nursery soil in a geographic area where few symbiotic fungi occur naturally. Soil fumigation obviously reduces populations of indigenous symbiotic fungi and other micro-organisms, improving the success of artificial soil inoculations with pure mycelial cultures of *P. tinctorius* and *T. terrestris*. Lastly, it appears that *C. graniforme*, owing to its inherently slow growth rate and its adaptation to drought-prone soils, will not effectively colonize roots of pine seedlings in irrigated nursery soils. Perhaps with maintenance of less soil moisture in nurseries where soil colonization by other symbiotic fungi is slow, this fungal symbiont may be effectively maintained on seedling roots.

Rate of mycelial inoculum.　During our research with *P. tinctorius* it became apparent that we did not know the least amount of inoculum needed to successfully infest soil and form ectomycorrhizae on seedlings. In many instances the amount of inoculum used in our early tests was dictated by the amount of inoculum available for use at the time. To examine this problem, a study was installed in a tree nursery in Mississippi (Marx, unpublished data). This nursery has been producing good quality pine seedlings for over 25 years that are usually heavily mycorrhizal with *T. terrestris* and other fungi. *Pisolithus* has also been observed in this nursery. In the spring of 1976, leached vermiculite–peat moss inoculum of *P. tinctorius* was broadcast on fumigated soil at rates of 2.80, 2.16, 1.62, 1.08, and 0.5 l/m^2 of soil surface. Two control treatments were installed; one received 2.8 l/m^2 rate of leached autoclaved inoculum and the other received no inoculum. These were used to delineate any possible physical or chemical effects of the inoculum to the soil and seedling growth. Seeds of *Pinus palustris* and *P. echinata* were planted and the plots mulched. In December, approximately 26 000 seedlings of each pine species were lifted and representative seedlings were evaluated.

Inoculation with *P. tinctorius* at any rate significantly increased total mycorrhizal development from 23 per cent (mean of both control groups) to 34 to 43 per cent on *P. palustris* seedlings. *Pisolithus* mycorrhizae, regardless of the inoculum rate, accounted for one-third of the total development. Significant increases in seedling growth and the number of plantable seedlings were associated with all inoculation treatments. Seedlings of *P. echinata* were also stimulated regardless of inoculum rate. On this species, *Pisolithus* mycorrhizae at the four highest inoculum rates, accounted for about one-third of the total development, but it formed only about one-quarter of all mycorrhizae at the 0.54 l/m^2 rate. *Thelephora terrestris* formed most of the other mycorrhizae. A comparison of seedlings and soil from the two different control treatments showed that the leached inoculum did not affect seedling growth or change chemical (major nutrients) or physical (cation exchange capacity) conditions of the soil. The 1.08 litres per m^2 of soil surface rate was the least amount that could effectively be used for maximum mycorrhizal development in this nursery. This nursery has one of the most rapid colonizations of fumigated soil by *T. terrestris* of any nursery in which we have worked. Therefore, this 1.08 l/m^2 rate may be even more effective in nurseries having a lesser degree of early competition from other fungi. We are currently recommending this rate for purposes of experimentation in properly fumigated nursery soils.

Drying of mycelial inoculum.   The weight and physical nature of pure mycelial inoculum that had been used up to this time caused certain problems. After leaching, the vermiculite–peat moss inoculum had a very high weight to volume index and physically resembled a sticky paste. The inoculum was also very heavy to transport and quite difficult to spread and mix into the soil. A study was conducted to determine the feasibility of drying the inoculum of *P. tinctorius* in order to eliminate these problems (Marx, unpublished data). Vermiculite–peat moss inoculum was leached in tap water, squeezed to remove excess water, placed 2 cm deep in an aluminium tray and dried to 12 per cent moisture at 28–30 °C for 56 hours in a forced-air oven. In order to ascertain the effects of drying on the efficiency of the inoculum, it was added to soil at different rates. Leached but non-dried inoculum was used in identical fashion for a further comparison. Inoculum was broadcast on fumigated soil in microplots at our nursery in Athens at rates of 2.16, 1.08, 0.54, and 0.27 l/m² of soil surface and mixed 10 cm deep into the soil. Control soil received vermiculite at the highest rate. Seed of *P. taeda* were planted in April 1976 and nearly 8000 seedlings were lifted and evaluated the following December. Dried inoculum was as good as, if not better than, non-dried inoculum for development of *P. tinctorius* mycorrhizae. An average of the three highest rates showed that *Pisolithus* from non-dried inoculum formed less than one-third of the 73 per cent total development while dried inoculum formed nearly half of the mycorrhizae. The lowest rate (0.27 l/m²) of both inoculum formulations formed less than half the amount of *Pisolithus* as the higher rates. Since procedures for soil fumigation are more efficient at our nursery facility than those employed in conventional nurseries, we obtained greater effectiveness at lower rates of both inoculum formulations in this study than obtained from the inoculum rate study in the Mississippi nursery. One reason for the greater effectiveness of the dried inoculum is that it mixes more homogeneously in soil than the paste-like, non-dried inoculum. Removal of excess water from leached inoculum reduced the volume by one-third; drying reduced it further by nearly a third. An initial 3 litre volume of inoculum from culture vessels is reduced to 1.2 to 1.4 litres of usable inoculum after it is leached and dried.

Storage of mycelial inoculum.   Severe limitations would be placed on the broad scale use of this type of inoculum of any fungus if the inoculum could not survive reasonable lengths of storage. There would be few cases where transport of inoculum from the laboratory to the nursery did not entail a period of storage under various temperature conditions. The following study (Marx, unpublished data) investigated the influence of length of storage at different temperatures on the effectiveness of dried and non-dried inoculum of *P. tinctorius*. Inoculum was prepared, as previously described, and 15 ml volumes were stored in test tubes at 5, 23, and 30 °C. At weekly intervals sets of tubes were removed from the incubators. The inoculum was mixed at a 1:8 ratio with fumigated soil and placed in small pots in the mycorrhizal fungus-free growth room. Seed of *P. taeda* were planted and seedlings were evaluated after four months. Non-stored, dried inoculum formed 50 per cent *Pisolithus* mycorrhizae and the non-dried inoculum formed 57 per cent. This proved initial viability of inoculum. After the first week of storage viability dropped to 48 per cent mycorrhizal development for non-dried and 41 per cent for dried inoculum. This level of viability was maintained for the next seven to nine weeks of storage for inoculum incubated at 5 and 23 °C and for five to seven weeks at 30 °C. Viability decreased significantly after longer periods. The fact that leached and dried inoculum of *P. tinctorius* can be stored for up to nine weeks at refrigeration temperatures and for at least five weeks at warmer temperatures indicates that it is quite durable and should withstand reasonable storage and transportation conditions.

It is apparent from the discussions of research carried out by scientists in various parts of the world that the artificial introduction of specific fungi into nurseries and containers is biologically feasible. Published reports indicate that inoculation programmes developed in Austria and Argentina are on a quasi-operational level for practical application. In the United States, a test programme is currently underway to determine the feasibility of producing inoculum of *P. tinctorius* for commercial uses. In the spring of 1977, Abbott Laboratories, Long Grove, Illinois, produced a dried, vermiculite-peat moss inoculum of *P. tinctorius* which we tested in 19 identical nursery experiments in 15 states of the South-East, South, and South-West. Seven different species of *Pinus* and *Quercus rubra* were involved. In each experiment different rates (1.62, 1.08, and 0.54 l/m² of soil surface) of the Abbott-produced inoculum and one rate (1.08 l/m²) of dried inoculum produced in our laboratory were compared. This inoculum was further evaluated in seedling container programmes in five different states involving eight species of *Pinus, Pseudotsuga menziesii,* and *Tsuga heterophylla.* From September 1977 to March 1978 over 150 000 seedlings were lifted and representative seedlings were evaluated in Athens. Although the results were erratic and somewhat inconsistent, they showed that the dried, vermiculite–peat moss inoculum of *P. tinctorius* can be produced in large volumes in industrial fermentors and is functional in forming mycorrizae on seedlings. In the spring of 1978 our tests were expanded to include the entire United States using an improved inoculum production method. 33 bare-root and 11 container nursery tests are currently underway. These tests involve all major ectomycorrhizal tree species grown in the United States. Studies will be terminated in the bare-root nurseries after one, two, or three years depending on the tree species under test. We are very optimistic about the biological value of this commercially produced inoculum. If this product form of *P. tinctorius* inoculum proves to be functional, Abbott Laboratories has the fermentor capacity to potentially produce hundreds of thousands of litres for application in world forestry.

Fungus selection criteria and maintenance of pure cultures

The most important first step in any nursery inoculation programme is the selection of the fungi (Bowen 1965; Marx 1977a; Mikola 1973; Moser 1963; Trappe 1977). The physiological differences that exist between mycorrhizal fungi can be used as criteria for their selection. The importance of each of the following criteria will vary according to the needs of the different inoculation programmes in different locations. Therefore, criteria will not be ranked in this discussion.

Host specificity

One criterion is host specificity. The consistent association of certain fungi for only a few specific tree hosts is well documented in the literature. Many other fungi are associated with a great number of different tree hosts (Marx 1977b; Stevens 1974; Trappe 1962). It is imperative, therefore, that the candidate fungi exhibit the physiological capacity to form mycorrhizae on the desired hosts. There is another aspect to this criteria, however. It is not sufficient to simply select a fungal species and then obtain an isolate for testing. Several isolates from different tree hosts and geographic regions should be used. This point has been stressed by Moser (1958c) and demonstrated by Theodorou and Bowen (1970) with isolates of *Rhizopogon luteolus*. We have obtained isolates of *P. tinctorius* from different species of oaks and compared them with isolates from pine in the mycorrhizal fungus-free room and in the microplot nursery on *Quercus rubra* seedlings (Marx, unpublished data). The pine isolates formed abundant mycorrhizae in the growth room and nursery. Some oak isolates formed a few mycorrhizae; some isolates did not form mycorrhizae at all. All isolates were similar in age and had comparable pigmentation and rates of vegetative growth in agar medium.

Growth in pure culture

Another criterion is the ability of the selected fungi to grow in pure culture; many ectomycorrhizal fungi will not. A variety of culture media (Moser 1958b; Stevens 1974; Trappe 1962) and methods of isolation (Palmer 1971) can be used to obtain pure cultures of the selected fungi. Ideally, the fungi should be able to grow rapidly (Moser 1959). Once cultures of the selected fungus have been obtained they must be maintained in a viable condition. Takacs (1967) recommends subculturing the stock cultures of the fungi every 60 days in order to retain vigour. Moser (1958b) stressed the need to subculture every two to five weeks, depending on fungus species, especially those to be used in current inoculation programmes. If the cultures are not subcultured frequently they exhibit poor growth and loss of pigment. He recommends growing declining cultures on a different medium to rejuvenate them. We found that continuous culturing of certain fungi on agar media for several years frequently decreased mycelial growth rate and the capacity to form mycorrhizae on pine. Changes in adaptive enzyme systems during continuous vegetative growth on synthetic medium probably accounts for this loss of ability to symbiotically infect the host roots. Mycelial agar discs cut from plate cultures and stored in sterile distilled water at 5 °C can be held for up to three years without loss of these physiological traits (Marx and Daniel 1976). The technique does not work for all fungi but is worthy of testing. The storage of cultures in a dormant physiological state should reduce, if not eliminate, shifts in adaptive enzyme systems. A certain amount of caution, therefore, must be used in evaluating fungi maintained in continuously growing, pure cultures for extended periods of time. We had isolates of *P. tinctorius* grown in continuous culture for 15 years that were still highly pigmented and grew at rates comparable to that achieved shortly after their isolation from sporophores. In 1974 these isolates lost their capacity to form mycorrhizae on pine. One of these (isolate 29) had been used successfully by us in several earlier studies (Marx, Bryan, and Davey 1970) since its original isolation. We have found, however, that our best isolates of *P. tinctorius* are those that are cycled back through their host every year or two and then reisolated from sporophores or directly from mycorrhizae. Our main isolate of *P. tinctorius* currently under test with Abbott Laboratories was first isolated in 1967 from a sporophore under a mature *P. taeda* growing in Georgia. This isolate has been rejuvenated by cycling it through pine hosts every one or two years. Today, it grows faster and forms more mycorrhizae on pine and oak than it did in 1967. It also has formed mycorrhizae on a variety of host species that other recently cultured isolates have not. A parent culture of this isolate maintained in continuous culture will form few mycorrhizae at this time.

If spores of a selected fungus are to be used for inoculum, then the ability of this fungus to grow well in pure culture is of little importance. However, growth in pure culture may be useful if it is to be reisolated from mycorrhiza to confirm its identity.

Once the growth potential of a fungus has been confirmed it is important to confirm its capacity to withstand physical manipulation (leaching, drying, soil incorporation, colonization by saprophytes, etc.). Producing large quantities of inoculum of a fungus is of little value if the fungus cannot survive the rigours of various manipulations

essential to inoculation of soil. Certain fungi grow readily in vermiculite–peat moss medium but cannot survive the leaching procedure or
soil inoculation. If we had not studied the colonization of non-
leached mycelial inoculum of *P. tinctorius* by various saprophytic
micro-organisms and rectified the problem by leaching, we could
have easily concluded that *P. tinctorius* was not a good fungal candidate for any inoculation programme.

Fungus adaptability

Another criterion is the adaptation of the selected fungus to the
major type of site on which the seedlings are to be outplanted. Of
equal importance is the ability of the fungus to survive and grow
under cultural conditions used in nurseries. According to Trappe
(1977), the ecological adaptability of an ectomycorrhizal fungus
hinges on the metabolic pathways it has evolved to contend with
environmental variation. Extremes of soil and climatic factors,
antagonism from other soil organisms, pesticide application, physical
disruption of mycelium from nursery operation, and the abrupt
adjustment from a fertilized and irrigated nursery soil to an unculti-
vated planting site with all of its stresses are only a few of the environmental variations to which the selected fungi must adapt.

The effect of temperature on different species and ecotypes of
ectomycorrhizal fungi is perhaps the most widely researched environmental factor. Upper and lower temperature limits of the candidate
fungi should be determined. Moser (1958d) studied the ability of
fungi to survive long periods (up to four months) of freezing ($-12\,^\circ$C)
and to grow at low temperatures (0–$5\,^\circ$C). He found that high elevation ecotypes of *Suillus variegatus* were not damaged after freezing
for two months, but valley ecotypes were killed after freezing for
only five days. Although not as striking, similar results were reported
with *S. tridentinus, S. plorans,* and *Gomphidius rutilus.* In low temperature growth studies, none of the species of *Amanita* grew at 5 or
$0\,^\circ$C. An interesting observation was that certain species and ecotypes which survived freezing for extended periods did not grow at
low temperatures. Generally, he found that mountain ecotypes and
species had much lower temperature optima than lowland ones. Even
after several years in pure culture at 20 to $23\,^\circ$C, the low temperature
fungi still maintained optima near $15\,^\circ$C. *Pisolithus tinctorius* not
only survives and grows well at unusually high temperatures, but also
it grows at $7\,^\circ$C and survives in frozen soil (Marx, Bryan, and Davey
1970, Marx and Bryan 1971, 1975). High temperature tolerance
makes *P. tinctorius* an excellent candidate for testing in the tropics
(Momoh and Gbadegesin 1975). *Rhizopogon luteolus* apparently is
not suitable for inoculation programmes because of its inability to

survive or grow at the high soil temperatures common to this area (Momoh 1973).

Reaction of the candidate fungi to soil moisture, organic matter, and pH are also important traits to consider. *Cenococcum graniforme* is not only drought tolerant but forms mycorrhizae in natural soils ranging in pH from 3.4 to 7.5 (Trappe 1964). We have observed *Pisolithus* mycorrhizae on pine in drought-prone coal spoils ranging in pH from 2.6 to as high as 8.4. Trappe (1977) has observed several species of fungi which form ectomycorrhizae in well-rotted conifer logs with a pH of 4.0 or lower in the Pacific North-west of the United States. Levisohn (1965) observed in England that *Suillus bovinus*, an excellent mycorrhizal fungus on spruce, naturally occurs in nursery soils containing abundant organic matter. Unfortunately the fungus disappears from the roots of spruce planted on sites having low organic matter. Its potential value in inoculation programmes would appear to be restricted to sites with high levels of organic matter.

Value of hyphal strands

Another criterion by which candidate fungi should be evaluated is their capacity to form hyphal strands in pure cultures and in soil. Bowen (1973) showed that nutrient uptake, especially phosphorus, is greater in fungi that produce hyphal strands. In Australia, one of the initial criteria for selection of fungi is their ability to produce hyphal strands under a wide range of conditions. Although research data is lacking we believe that the abundant hyphal strands produced by *P. tinctorius* not only enhance nutrient absorption, but increase its survival potential under adverse conditions. Yellow-gold hyphal strands of *P. tinctorius*, easily visible to the naked eye, have been traced through highly toxic and hot coal spoils as far as 4 m from seedlings to sporophores by Schramm (1966) and others (Marx 1977a). On an exposed borrow pit in South Carolina we traced hyphal strands of *P. tinctorius* over 3 m from mycorrhizal roots of *P. palustris* to sporophores.

Aggressiveness of fungus

Another extremely important criterion is the aggressiveness of the candidate fungus to feeder roots. The fungus should have the capacity to form abundant mycorrhizae as soon as feeder roots are formed. It must be able to maintain superiority over naturally occurring fungi in the nursery. Aggressiveness is best evaluated by making quantitative assessments on the amount of mycorrhizae formed by the introduced fungus at different intervals of time. Quantitative assessments are the only valid parameter which can be used to judge the effectiveness of inoculations. We have found that maximum benefit of *Pisolithus* mycorrhizae is achieved on pine seedlings when at least two-thirds of all the mycorrhizae on the seedlings are formed by *P. tinctorius*.

Field performance of seedlings with specific ectomycorrhizae

The ultimate proof of the value of inoculation of bare-root or container grown nursery seedlings with specific fungi is their performance under diverse field conditions. Meaningful conclusions can only be obtained from properly designed, installed, and maintained field experiments which include periodic tree measurements and mycorrhizal assessments conducted over several years. Only limited field data of this type is available in the literature. Moser (1963) reported that spruce seedlings with mycorrhizae formed by *Phlegmacium glaucopus* survived and grew better than comparable non-mycorrhizal seedlings on a 2100 m altitude forest site in Austria. In another test, four-year-old nursery grown seedlings of *Pinus cembra* with few mycorrhizae were planted on a 2100 m altitude site. These seedlings were inoculated (apparently at planting time) with an equal mixture of pure mycelial inoculum of *Suillus plorans, S. placidus, Paxillus involutus,* and *Amanita muscaria*, a mixture of mycelial inoculum of these four mycorrhizal fungi contaminated with *Penicillium, Mucor,* and bacteria, or no inoculum. After three years the mixed inoculum (either pure or half-pure) of the symbiotic fungi stimulated height growth and increased the number of healthy seedlings by 65 per cent over the non-inoculated controls. Half-pure inoculum of the fungi was only slightly less effective in stimulating seedling survival and growth than the pure mycelial mixture. An assessment of mycorrhizal development was not reported in this study.

Puerto Rico

In Puerto Rico in 1965, Vozzo and Hacskaylo (1971) outplanted seedlings of *Pinus caribaea* from one of the container nursery experiments on a sandy loam site. Unfortunately, damage to seedlings by vandals and cattle shortly after planting resulted in study termination after only six months. Results showed, however, that mycorrhizae formed by *Suillus cothurnatus, Rhizopogon roseolus, Corticium bicolor,* and by unidentified fungi in natural soil inoculum stimulated height growth of the seedlings over both fertilized and non-fertilized, non-mycorrhizal seedlings. Regardless of fertility, non-mycorrhizal seedlings were chlorotic and stunted. At the time of planting, 75 per cent of the seedlings inoculated with pure mycelial cultures and 95 per cent of the seedlings inoculated with the natural inoculum had ectomycorrhizae. Apparently the degree of development on individual seedlings was not determined. Their results indicated that *S. co-thurnatus, R. roseolus,* and *C. bicolor* in pure mycelial inoculum can be used in Puerto Rico for the establishment of *P. caribaea.*

Australia

In Australia, Theodorou and Bowen (1970) installed two field experiments with *Pinus radiata* seedlings. In the first test, one-week-old seedlings were transplanted into fumigated potting mixture contained in small wooden veneer tubes. Each tube contained a 10 g layer (3 cm deep in tubes) of pure mycelial inoculum of either *Suillus granulatus, S. luteus,* or two isolates of *Rhizopogon luteolus.* Sterile medium was added to control tubes. The seedlings were grown for four months in a greenhouse and then hardened off for an additional three months in the open prior to outplanting. The seedlings were planted in 1966 on a loamy soil field site some 200 m from an established stand of *P. radiata.* At planting, all inoculated seedlings were 7 cm tall and had about 25 per cent mycorrhizal development. Control seedlings were 6 cm tall with only a trace of mycorrhizae. The field design was a randomized design with three blocks. A buffer row surrounded each plot within each block. Significant differences in height occurred as early as six months after planting. Seedlings with *S. granulatus* or *R. luteolus* mycorrhizae were about 46 per cent taller (13.9 cm) than control seedlings (9.5 cm). Seedlings with *S. granulatus* were also noticeably greener than seedlings of other treatments. All seedlings had a healthy green colour after 28 months. Following a summer drought, nearly three times more control seedlings had died (13 per cent) than did those inoculated with *S. granulatus* or *R. luteolus* (3–5 per cent). 20 per cent of the seedlings with *S. luteus* died during the drought. All inoculated seedlings were significantly taller than controls after eight months. After 32 months, the rate of height growth of seedlings with *S. granulatus* mycorrhizae was significantly greater than control seedlings. At 36 months the rate of growth was similar, but differences in height that developed earlier were still evident. Root evaluations revealed that differences in growth due to the different fungi were related to the degree of ectomycorrhizal development. *Suillus granulatus* formed significantly more mycorrhizae (81 per cent) during the 36 months than did the two *R. luteolus* isolates (78 per cent), *S. luteus* (68 per cent), or the control seedlings (65 per cent). White mycorrhizae typical of those produced by the test fungi dominated inoculated seedling roots and a brown-type was observed on the control seedlings.

In their second field test Theodorou and Bowen (1970) grew *Pinus radiata* seedlings as before, except *S. granulatus, S. luteus, R. luteolus,* and an unidentified isolate obtained from mycorrhizae of nursery seedlings were used. At planting all seedlings were 6 to 7 cm tall. Inoculated seedlings had 16 to 23 per cent mycorrhizal development and control seedlings had none. These seedlings were outplanted 900 m from an established *P. radiata* stand. After 23 months, there were no significant differences in seedling heights or development of mycorrhizae between treatments. The control seedlings had a similar amount of mycorrhizae (76 per cent) to the inoculated seedlings (69 to 82 per cent). A white ectomycorrhizal type which apparently occurred naturally on this site was observed early on inoculated and control seedlings. This natural colonization of roots of control seedlings obviously minimized the effect of inoculations. The authors stressed the need for larger field plots, more extensive buffers between plots, and test sites which do not contain ectomycorrhizal fungi of *P. radiata* for future studies. They feel these conditions are essential to valid testing of the significance of the various fungi. In spite of problems in the second study, their results suggested that pure mycelial cultures of *S. granulatus* and *R. luteolus* can be used to form abundant mycorrhizae and generally improve field performance of *P. radiata* seedlings. In the first test, *S. luteus* improved field performance but to a lesser degree.

Adverse sites in United States

Beginning in 1973, field tests were conducted in the United States to ascertain the value of mycorrhizae formed by *Pisolithus tinctorius* and other fungi for improving survival and growth of pines on adverse sites. Since most of this data was summarized recently (Marx 1977a), only a few examples will be briefly discussed and updated here. One of our first tests was installed in Kentucky on a very toxic (pH 3.8) coal spoil that had been unsuccessfully planted with pine seedlings several times. Seedlings of *Pinus virginiana* were produced in our nursery with *Pisolithus* and *Thelephora* mycorrhizae using methods described earlier (Marx and Bryan 1975). The seedlings were graded to similar heights and root collar diameters; all had about 75 per cent mycorrhizal development. Seedlings inoculated with *P. tinctorius* had two-thirds of this amount formed by *P. tinctorius*. All other mycorrhizae were formed by *T. terrestris*. The field design was random with five blocks. Test plots within each block were separated by a 4 m non-planted border. After two years only two of the 160 seedlings with *Thelephora* mycorrhiza survived, whereas 78 of the 160 seedlings with *Pisolithus* mycorrhiza survived. More significant was the growth of *Pisolithus* seedlings which produced an average seedling volume* of 130 cm^3 compared to the two *Thelephora* seedlings with an average of 3 cm^3. This volume of

*Seedling volume (cm^3) = (root collar diameter, cm)^2 $\times$ height, cm.

Thelephora seedlings was the same as that measured at planting, indicating that these seedlings did not grow during this two-year period.

A similar planting with five blocks was installed on a coal spoil (pH 3.4) in Virginia with *P. taeda* seedlings produced as before. Unfortunately, trees destroyed by vandals precluded accurate assessments of survival. Growth measurements after two years showed that seedlings with *Pisolithus* mycorrhiza has an average seedling volume of 962 cm^3 and those with *Thelephora* mycorrhiza had a volume of 379 cm^3. The last example of a coal spoil study was installed on another toxic (pH 3.9) site in Kentucky. This spoil was unsuccessfully planted twice with nursery seedlings of *P. taeda*. Seedlings of *P. taeda* and *P. echinata* were produced as before with *Pisolithus* and *Thelephora* mycorrhizae and graded to similar sizes and ectomycorrhizal development. The field design was randomized with five blocks. After two years survival was not influenced by treatments, but growth was strongly affected. *Pinus taeda* seedlings with *Pisolithus* mycorrhiza had plot volume indices* of 13 000 cm^3 and those with *Thelephora* had 2000 cm^3. *Pinus echinata* seedlings with *Pisolithus* mycorrhiza had plot volumes of 3600 cm^3 compared to *Thelephora* seedlings with a volume of 700 cm^3 (Marx and Artman 1979).

We prefer this plot volume index (PVI) parameter because it integrates survival, height, and root collar diameters into a single value for comparison. It also represents an accurate measure of the response of all seedlings to treatment. We have found that height measurements alone give poor representations of pine seedling performance.

In all the field tests on coal spoils yearly root evaluations were made. The results confirmed that *P. tinctorius* is ecologically adapted to these harsh sites. Without exception, seedlings with *Pisolithus* mycorrhiza at planting had new roots totally colonized by this fungus. Colonization was so prolific that after the second year hyphal strands were easily detected in the spoil material with the unaided eye as far as 3 m from the young seedlings. Production of sporophores of *P. tinctorius* in the test plots was also prolific. In one *Pisolithus* plot (49 m^2) of *P. taeda*, 83 sporophores were collected during one visit. In contrast, on seedlings that initially had *Thelephora* mycorrhizae new root growth was minimal and only a few were colonized by *Thelephora* at the end of the first year. Many of the original *Thelephora* mycorrhizae (mycorrhizae located on the original root system) were necrotic and only a few visually appeared to be functional. After the second year, a low incidence (three to five per cent) of *Pisolithus* mycorrhizae was detected on newly formed roots on one site; *Thelephora* had spread slowly to new roots on the two other

*Plot volume index, cm^3 (PVI) = (root collar diameter, cm)^2 × height, cm × No. surviving tree seedlings per plot.

sites. Sporophores of *T. terrestris* were only occasionally observed in *Thelephora* plots.

Fourth year data were recently collected and evaluated from the tests on coal spoils in Virginia and the last site in Kentucky. Differences in growth were greater after four years than those measured after two years. One interesting observation was made after the severe winter of 1976-7. Winter scorch of needles was severe on all seedlings on both sites. Seedlings with *Pisolithus* mycorrhiza, however, recovered from this severe needle browning at least six weeks earlier in the spring than seedlings with *Thelephora* mycorrhiza. This earlier recovery undoubtedly gave seedlings with *Pisolithus* mycorrhiza a longer active growth period than seedlings with *Thelephora* mycorrhiza.

A variety of field tests have also been installed on other types of adverse sites (Marx 1977a). Only a few of these will be discussed here. Kaolin spoils are created by strip mining. They are nutrient deficient, reflect sunlight from their light colour, and usually are drought-prone and highly compacted, but most lack the toxic characteristics of coal spoils. In 1975, *P. taeda* seedlings with mycorrhizae formed by either *P. tinctorius, T. terrestris,* or *C. graniforme* were produced in our mycorrhizal fungus-free growth room by Otrosina (1977). Seedlings inoculated with *P. tinctorius* had 85 per cent *Pisolithus* mycorrhizal development. Those inoculated with *Cenococcum* had 20 per cent *Cenococcum* and 30 per cent *Thelephora* (contaminant) mycorrhizae. Seedlings inoculated with *T. terrestris* had 85 per cent *Thelephora* mycorrhizae. These seedlings were outplanted in central Georgia on two different kaolin spoils. Both sites were covered with 7.5 cm of forest soil prior to planting. This is a recommended practice to assist in reclamation of the spoils. Treatments were arranged in a random block design with four blocks. Half of the seedlings were fertilized with 170 g of 10:10:10 NPK fertilizer per seedling at planting and the other half received no fertilizer. Only first year data are available. On one site, seedlings with *Cenococcum* mycorrhizae survived better in the fertilized (83 per cent) than in the non-fertilized (55 per cent) plots, but seedling growth was similar. Survival (65 and 73 per cent) and growth of seedlings with *Thelephora* mycorrhizae were not affected by fertilization. Generally, seedlings with *Cenococcum* grew better than seedlings with *Thelephora* mycorrhiza. In the non-fertilized plots, seedlings with *Pisolithus* mycorrhiza were significantly larger and survived better (73 per cent). Overall seedling growth on the second site was generally greater than growth on the first site. Fertilization significantly improved height growth but specific mycorrhiza had no effect (only *Pisolithus* and *Cenococcum* tested). Root evaluation at the end of the growing season revealed that all introduced fungi persisted on roots of their respective seedlings. However, on the first site the degree of mycorrhiza developed by each fungus decreased; on the second site, the degree of development was greater for both *Pisolithus* and *Cenococcum* than on the other site. This author also concluded that ectomycorrhizal fungi, such as *P. tinctorius* and *C. graniforme*, ecologically adapted to adverse soil conditions, afford improved survival and growth to pine seedlings on kaolin spoils over seedlings with mycorrhizae formed by fungi such as *T. terrestris*. In earlier work on kaolin spoils (Marx 1977a), hyphal strands of *Pisolithus* were found to spread very rapidly through the spoil material. In one study, spread was so rapid that by the end of the first growing season hyphal strands had grown nearly 2 m from inoculated seedlings and had formed mycorrhizae on control seedlings planted in adjacent rows. The integrity, therefore, of the ectomycorrhizal treatments was lost. This rapid spread between rows brought about changes in subsequent experimental designs. Since 1974, we plant seedlings with different mycorrhizal fungi in discrete plots separated by at least a 2 m non-planted border. Frequently, if space is available, a 4 m strip is left non-planted.

Severely eroded sections of the Copper Basin of Tennessee were also used as outplanting sites. Since the 1840s thousands of hectares of productive forest were decimated by cutting timber and using the wood in heap roasting of mineral ores. This roasting produced SO_2 which further damaged vegetation. Severe sheet erosion followed this destruction of vegetation. Air quality was improved in 1964 and reforestation efforts began, but with little success. The surface soil in most places in the basin is gone, leaving nothing more than exposed parent material. The soil has low levels of available nutrients, high temperatures during summer months, and poor internal water drainage, but does not contain levels of any element toxic to pines. Root examinations of pines planted in the basin reveal two obvious forms of ectomycorrhizae. *Thelephora* mycorrhizae and its sporophores occur sporadically. It was probably introduced on roots of the outplanted seedlings from a nursery. The other type, formed by *Pisolithus*, occurs naturally on pines and oaks on the perimeter of the Basin and was probably introduced by wind-borne basidiospores.

Tests were installed at two locations in the Basin in March 1974 (Berry and Marx 1978). Seedlings of *Pinus taeda* and *P. virginiana* were produced in our microplot nursery with *Pisolithus* and *Thelephora* mycorrhizae as described earlier. All seedlings were graded to similar sizes and degrees of mycorrhizal development. Those inoculated with *Pisolithus* had over two-thirds of the total of 70 per cent mycorrhizal development formed by *Pisolithus* and seedlings inoculated with *Thelephora* had about 70 per cent development by *Thele-phora*. Prior to planting, the test sites were levelled and the soil was ruptured (subsoiled) to a depth of 60 cm to destroy hardpans and to allow roots and water to more readily penetrate the soil. Fertilizer (672 kg/ha of 10:10:10 NPK) and dolomitic limestone (4480 kg/ha) were broadcast and disced into the soil of all plots. The trees were planted in a randomized design with three blocks. Each plot was surrounded by a border row of trees and a 2 m non-planted strip. After two years, survival of both pine species (88 to 99 per cent) was not affected by mycorrhizal treatments. *Pisolithus* mycor-rhizae significantly increased height and stem diameters of both pine species on one site. Seedling volumes of *P. taeda* were 93 per cent and *P. virginiana* were 90 per cent greater than those of comparable seedlings with *Thelephora* mycorrhiza. Unfortunately excessive ex-perimental variations precluded statistical significance on the second site, even though *Pisolithus* mycorrhiza increased seedling volume by 45 per cent for *P. taeda* and 26 per cent for *P. virginiana* seedlings. Both fungi persisted well on roots, since root evaluations showed that they were still dominant on their respective seedlings. Sporo-phores of *P. tinctorius* were also produced in great abundance in the *Pisolithus* plots.

The last example of research done on adverse sites is from a bor-row pit in the lower piedmont of South Carolina (Ruehle, unpub-lished data). A borrow pit is an area from which soil was removed (borrowed) for use in construction. This eight hectare site was created in the 1950s by vertical removal of from 1 to 5 m of soil. The result-ing surface material had physical and chemical characteristics very similar to those of kaolin spoils. In 1955, the site was planted with nursery seedlings of *P. taeda*. By 1975, most of the trees were less than 2 m tall and very chlorotic; root penetration was very restricted. *Pisolithus* occurred naturally on many of these trees. Prior to study installation in 1975, the trees were removed and the site was levelled. The area was then subsoiled as described earlier. Fertilizers (560 kg/ha of 10:10:10 NPK) and dolomitic limestone (2240 kg/ha) were broad-cast on half the plots and the other half received a 1 cm layer of dried sewage sludge. All plots were disced and seeded to grass. In 1976, seedlings of *P. taeda* were grown in styrofoam containers with vermiculite–peat moss substrate. Prior to seeding, the substrate in one-third of the containers was mixed 8:1 with leached, non-dried pure mycelial inoculum of *P. tinctorius*. These seedlings were grown in the greenhouse. One-third were grown in the greenhouse without artificial inoculation for natural colonization by *T. terrestris*. The remaining one-third were grown in the mycorrhizal fungus-free growth room in a non-mycorrhizal condition. All seedlings were watered as needed and fertilized lightly. After four months, all seed-lings, regardless of treatment, were about 10 cm tall. Those inoculated with *Pisolithus* had about 20 per cent mycorrhizae formed by *Pisolithus* and 30 per cent mycorrhizae formed by naturally occurring *T. terrestris*. The second group had a total of 65 per cent mycorrhizal development, all formed by *T. terrestris*. The third group from the growth room lacked mycorrhizae. Seedlings were planted in November 1976 in a randomized design with five blocks. Each plot was surrounded by a border row of seedlings and a 4 m non-planted space.

Only data from one growing season are available. In the sludge amended plots, survival was 90 per cent for pine seedlings with a mixture of *Pisolithus* and *Thelephora* mycorrhiza, 75 per cent for those with just *Thelephora* mycorrhiza, and only 62 per cent for non-mycorrhizal seedlings. The PVIs were 1702 cm^3 for seedlings with *Pisolithus* mycorrhiza, 361 cm^3 for those with *Thelephora*, and only 104 cm^3 for those without mycorrhiza. The differences due to *Pisolithus* mycorrhiza in comparison to *Thelephora* and non-mycorrhizal seedlings represent increases of 372 and 153 per cent, respectively. In the fertilized plots similar results were obtained but they were not as striking. Survival was 98, 89, and 88 per cent, respectively, for seedlings with *Pisolithus*, *Thelephora*, or no mycorrhizae. The corresponding PVIs were 75, 61, and 27 cm^3. Root evaluation data are not available at this time.

These results not only show that seedlings with mycorrhizae survive and grow better than seedlings lacking mycorrhizae, but they also show that seedlings with *Pisolithus* mycorrhiza are better adapted to adverse soils in borrow pits even after amendments with sludge or fertilizers.

We can conclude from these field studies on adverse sites that reclamation and reforestation of such sites can be expedited by using pine seedlings tailored with mycorrhizae formed by fungi capable of growing under adverse conditions. Thus, the planting of seedlings with root systems physiologically and ecologically adapted to accommodate the adversities of the planting site can be an important biological tool in reforestation. In reality, however, sites such as coal and kaolin spoils and borrow pits are not the only adverse sites created by the activities of man. Many reforestation sites, especially those which have been intensively site prepared (stump shearing, root raking, slash removal, burning, discing, etc.), are temporarily adverse (Schultz 1977). Until vegetation is re-established by either natural or artificial means, the mineral soils are often exposed and subject to broad fluctuations of temperature, moisture, and fertility, as well as to erosion and compaction (Haines, Maki, and Sanderford 1975).

These are adverse soil conditions, however temporary, to which root systems of newly planted seedlings will be exposed. If these soil conditions are extreme, the survival and early growth of seedlings with mycorrhizae formed by nursery adapted fungi, i.e. *Thelephora*, may be unduly affected. This point could explain certain reforestation failures in the past on sites which have been intensively prepared.

Routine sites in United States

After considering these points, we installed several studies on routine reforestation sites to compare the effects of *Pisolithus* and *Thelephora* mycorrhiza on establishment and early growth of various pines. Since 1974, nearly 75 000 seedlings have been experimentally outplanted on a variety of reforestation sites. In most studies, seedlings were graded so that different amounts of *Pisolithus* and *Thelephora* mycorrhizae were treatment variables. The following are results of one such study (Marx, Bryan, and Cordell 1977). Pine seedlings were produced in Florida and North Carolina nurseries with *Pisolithus* and naturally occurring *Thelephora* mycorrhizae. The nursery phase of this study was described earlier (Marx *et al.* 1976). The use of basidiospores and pure mycelial inoculum of *P. tinctorius* resulted in seedlings with different quantities of *Pisolithus* mycorrhizae. In the Florida nursery, seedlings of *P. taeda, P. elliottii* var. *elliottii,* and *P. clausa* were graded to equal size and to 65 per cent mycorrhizal development. Those inoculated with mycelial inoculum had three-quarters of this 65 per cent formed by *Pisolithus* and those inoculated with basidiospores had about one-third of this amount formed by *Pisolithus*. Remaining mycorrhizae on these seedlings and the control seedlings were formed mainly by *Thelephora terrestris*. Three sites in Florida, recently clearcut of pines, were site prepared and all slash was burned exposing mineral soil. One site was a deep sand ridge; one was a palmetto flatwood. The third site was also a flatwood site but it was planted to only *P. elliottii* var. *elliottii* with abundant *Pisolithus* or *Thelephora* mycorrhizae. No intermediate amount of *Pisolithus* was used. This study also involved a fertility variable of 90 kg/ha of both nitrogen and phosphorus.

In the North Carolina nursery, seedlings of *P. taeda, P. virginiana,* and *P. strobus* were graded to equal size and also to 75 per cent mycorrhizal development. Generally, seedlings inoculated with mycelial inoculum had at least eight-tenths of the 75 per cent total development formed by *Pisolithus* and those inoculated with basidiospores had between one-tenth to two-thirds of the total mycorrhizae formed by *Pisolithus*. The two test sites in North Carolina were cleared of pine and oak, site prepared, and all slash was piled and burned. One site was considered a good reforestation site because of the presence of 25 cm of top soil. The other was considered poor because it was an eroded slope.

Seedlings on four sites were planted in a random design with five blocks. The fertilizer–mycorrhiza study had only three blocks. Plots within blocks were separated by at least a 3 m non-planted strip.

After two years, seedlings with the greatest quantity of *Pisolithus* mycorrhizae at planting generally survived and grew better than seedlings with the same amount of mycorrhizae formed by *T. terrestris*. Seedlings with less *Pisolithus* mycorrhizae were usually intermediate between these two seedling groups in survival and growth. In North Carolina, abundant *Pisolithus* mycorrhizae significantly increased PVI of both *P. virginiana* and *P. taeda* by about 25 per cent on the good site and by about 50 per cent on the poor site in comparison to seedlings with only *Thelephora* mycorrhizae. *Pinus taeda* seedlings with less *Pisolithus* and more *Thelephora* mycorrhizae (from basidiospores) on the better site had a 30 per cent greater PVI than did the *Thelephora* seedlings; they were not different on the poor site. Greater differences occurred on seedlings of *P. strobus*. Those with abundant *Pisolithus* mycorrhizae had five times greater PVI than seedlings with *Thelephora*. This test with *P. strobus* is somewhat unusual since these seedlings were outplanted after only one growing season in the nursery instead of the normal two growing seasons.

In Florida, the results were similar. On both sites *P. taeda* and *P. elliottii* var. *elliottii* seedlings with abundant *Pisolithus* mycorrhizae had significantly greater PVI (16 to 51 per cent) than *Thelephora* seedlings. On the palmetto flatwood, *P. taeda* seedlings with only one-third of the complement of mycorrhizae formed by *Pisolithus* also had a PVI that was 46 per cent greater than *Thelephora* seedlings. The intermediate amount of *Pisolithus* on seedlings of *P. elliottii* var. *elliottii* did not have an effect on growth. The most significant results were obtained with *P. clausa*. Seedlings with abundant *Pisolithus* had PVIs on both sites that were from 270 to 445 per cent greater than seedlings with *Thelephora*. The greatest difference in survival and growth occurred on the flatwood site which is considered off-site for *P. clausa*. In the fertility study, *P. elliottii* var. *elliottii* seedlings with *Pisolithus* mycorrhiza in the fertilized plots were the same size as seedlings with *Thelephora*. However in the non-fertilized plots, seedlings with *Pisolithus* mycorrhiza had a PVI that was 175 per cent greater than comparable seedlings with *Thelephora* mycorrhiza. Yearly root evaluation for two years of representative seedlings on the various sites revealed that *Pisolithus* persisted well, especially on those sites where it stimulated the most seedling growth. On all sites, particularly the better ones, other ectomycorrhizal fungi such as *Cenococcum graniforme* or an unidentified *Rhizopogon* were

Fig. 2.1. Growth response of pines to specific ectomycorrhizae. Upper left and right photographs are loblolly and Virginia pine seedlings after two years on a routine reforestation site in North Carolina, USA, that had abundant *Pisolithus tinctorius* mycorrhizae on roots at planting time. Lower left and right photographs are corresponding control seedlings that had abundant naturally occurring *Thelephora terrestris* mycorrhizae on roots at planting time. Horizontal lines in background are spaced 10 cm.

observed on new roots of various seedlings regardless of initial mycorrhizal conditions. Sporophores of *P. tinctorius* were also observed in many *Pisolithus* plots. We obtained fourth year data from these seedlings recently (Marx, unpublished data). The differences in growth rate are either greater after the fourth growing season or are approximately the same as those at the end of the second season.

Krugner (1976) grew seedlings of *P. taeda* in a nursery soil of different fertility inoculated with *P. tinctorius, T. terrestris,* or a mixture of the two fungi. As discussed earlier, *Pisolithus* in this study formed more mycorrhizae in fertilized than in non-fertilized soil. Selected seedlings were outplanted on two recently prepared sites in the coastal plain of North Carolina. Only data from the first growing season is available. On one site, excessive weed competition and high populations of indigenous ectomycorrhizal fungi apparently eliminated the effects of the inoculation and fertility treatments from the nursery. The other site was considered poor and seedlings with the greatest amount of *Pisolithus* mycorrhizae had up to 19 per cent better survival, 63 per cent more height growth, and 15 per cent larger stem diameters than seedlings with only *Thelephora* mycorrhiza.

These results indicate that under the temporarily adverse situations caused by tree removal and site preparation of routine reforestation sites, pine seedlings may survive and grow faster if they have abundant mycorrhizae formed by a fungus, such as *P. tinctorius*, which is ecologically adapted to adverse conditions. Apparently, fertilization reduces the adverse situation to such a degree that seedlings with *Thelephora* mycorrhizae can survive and grow as well as those with *Pisolithus*. Our results also show that the more *Pisolithus* mycorrhizae seedlings have on their roots at planting, the more benefit they derive from this specific mycorrhizal association, especially on the poorer reforestation sites. Perhaps the ecological adaptation to poor soil conditions allows *P. tinctorius* a competitive advantage over other ectomycorrhizal fungi for colonization of new feeder roots. On better reforestation sites the other fungi may be more competitive and aggressive than *Pisolithus*. We have found from other field experiments that seedlings with *Thelephora* mycorrhizae in non-stress situations survive as well and grow better than seedlings with *Pisolithus* mycorrhiza. These observations coincide very well with what we think we know about the biological significance of certain ecological adaptability traits of these fungi.

Results from these various field studies show that specific ectomycorrhizal fungi can improve initial field performance of tree seedlings on good and poor sites. Some fungi appear to increase tolerance of the seedlings to extremes in soil environment, whereas others appear to enhance absorption of certain nutrients, such as phosphorus, from the soil. In all probability, many of these fungi share certain traits which act in concert to increase survival or early growth of tree seedlings. Unfortunately, there is no data to show whether these early growth effects have any influence on the final volume of wood harvested at the end of the normal rotation. Only long-range studies can furnish information of this type.

Conclusion

There is no doubt that a variety of proven methods are available to ensure the development of ectomycorrhizae on forest tree seedlings for the establishment of man-made forests. Certain methods have more advantages than others. Some methods, such as the use of pure mycelial inoculum, have more biological advantages than others, but a great deal more research must be done. There is sufficient information, however, to conclude that pure cultures of certain fungi, such as *Suillus granulatus, Rhizopogon luteolus*, and *Pisolithus tinctorius* can be used to assure good survival and growth of tree seedlings on a variety of sites. These represent only the beginning of the practical concept, however. When one considers the millions of hectares of potential exotic forests which should be established in Third World nations, as well as the millions of hectares of former forest lands awaiting artificial regeneration throughout the world, the importance of the selection, propagation, manipulation, and management of superior strains or species of mycorrhizal fungi as a forest management tool is paramount. Research so far has only revealed the tip of the iceberg in regard to potential use in world forestry. There still remains a tremendous reservoir of basic and practical information which must be revealed if these fungi are to be managed and, therefore, fully utilized in forest regeneration.

The introduction of ectomycorrhizal fungi into various parts of the world to establish exotic forests has expanded the geographic range of these fungi throughout the world (Mikola 1969, 1970, 1973). Although many species probably died, numerous fungi are currently thriving in areas far distant from their original habitat. These fungi, either individually or as a group, have a tremendous capacity to adapt to different environments. Once techniques have been perfected for use of pure cultures and adequate quantities of inoculum are available, the specific fungi should be tested on tree seedlings over a wide range of environmental conditions encountered in forestation throughout the world. There is not sufficient information available in the world literature on the use of a specific fungus to even remotely suggest where it can or cannot benefit a specific tree species in a given locality. Even though the effect of a given fungus may only be temporary, its short term influence may make the difference between initial success or failure of seedling establishment. There are several botanical precedents for this idea of testing an organism over a spectrum of environmental conditions beyond those present in its natural habitat. One such example is *Pinus radiata*. Its natural range is restricted to about 4500 hectares along the coast of California. Since the mid-1800s this tree has been planted successfully throughout the world, including such countries as Australia, New Zealand, Chile, Bolivia, Spain, Ireland, and several African nations. By 1958, over 623 000 hectares had been planted (Scott 1960). In many of these countries it has become the major commercial forest tree. There is little doubt that these forests are established today because foresters took a broad ecological view of the potential range of *Pinus radiata*. Researchers on mycorrhizae should also approach the use of specific ectomycorrhizal fungi in world forestry from a broad ecological view until results from research dictate otherwise. Let us determine the biological and practical significance of a given ectomycorrhizal fungus to forest productivity not by supposition, but by facts obtained by using scientific rules of proof.

References

Balmer, W. E. (1974). Containerization in the Southeast. In *Proc. N. Amer. Containerized Forest Tree Seedling Symp.* (eds. R. W. Tinus, W. I. Stein, and W. E. Balmer) pp. 38–41. Great Plains Agric. Council Publ. No. 68.

Berry, C. R. and Marx, D. H. (1978). Effects of *Pisolithus tinctorius* ectomycorrhizae on growth of loblolly and Virginia pines in the Tennessee Copper Basin. USDA Forest Serv. Res. Note SE-264.

Bowen, G. D. (1962). Uptake of phosphate by mycorrhizas of *Pinus radiata*. *Third Australian Conf. in Soil Sci.* Canberra.

— (1965). Mycorrhiza inoculation in forestry practice. *Aust. For.* 29, 231–7.

— (1973). Mineral nutrition of ectomycorrhizae. In *Ectomycorrhizae: their ecology and physiology* (eds. G. C. Marks and T. T. Kozlowski) pp. 151–205. Academic Press, New York.

— and Theodorou, C. (1973). Growth of ectomycorrhizal fungi around seeds and roots. In *Ectomycorrhizae: their ecology and physiology* (eds. G. C. Marks and T. T. Kozlowski) pp. 107–50. Academic Press, New York.

Briscoe, C. B. (1959). Early results of mycorrhizal inoculation of pine in Puerto Rico. *Carib. Forester* 20, 73–7.

Chevalier, G. and Grente, J. (1973). Propagation de la mycorhization par la truffe a partir de racines excisees et de plantules inseminatrices. *Ann. Phytopathol.* 4, 317–18.

Clements, J. B. (1941). The introduction of pines into Nyasaland. *Nyasald agric. q. J.* 1, 5–15.

Donald, D. G. M. (1975). Mycorrhizal inoculation of pines. *Jl S. Afr. For. Ass.* 92, 27–9.

Ekwebelam, S. A. (1973). Studies of pine mycorrhizae at Ibadan. Res. Paper, For. Ser., No. 18. Fed. Dept. For. Res., Nigeria.

Fontana, A. and Bonfante, P. F. (1971). Sintesi micorrizica di *Tuber brumale* Vitt. con *Pinus nigra* Arnold. *Allionia* 17, 15–18.

Gerdemann, J. W. and Trappe, J. M. (1974). The Endogonaceae in the Pacific Northwest. *Mycol. Mem.* 5, 1–76.

Gibson, I. A. S. (1963). Eine Mitteilung über die Kiefernmykorrhiza in den Wäldern Kenias. In *Mykorrhiza* (eds. W. Rawald and H. Lyr) pp. 49–51. Fischer, Jena.

Göbl, F. (1975). Erfahrungen bei der Anzucht von Mykorrhiza-Impfmaterial. *Cbl. Gesamte Forstwesen.* 92, 227–37.

Goss, R. W. (1960). Mycorrhizae of ponderosa pine in Nebraska grassland soils. Univ. of Nebraska, College of Agric. Res. Bull. No. 192.

Grand, L. F. (1976). Distribution, plant associates and variations in basidiocarps of *Pisolithus tinctorius* in the United States. *Mycologia* 68, 672–8.

Hacskaylo, E. (1965). *Thelephora terrestris* and mycorrhiza of Virginia pine. *Forest Sci.* 11, 401–4.

— (1971). Metabolic exchanges in ectomycorrhizae. In *Mycorrhizae* (ed. E. Hackskaylo), USDS Forest Serv. Misc. Publ. No. 1189, pp. 175–96.

— and Palmer, J. G. (1957). Effects of several biocides on growth of seedling pines and incidence of mycorrhizae in field plots. *Pl. Dis. Reptr.* 41, 354–8.

— and Vozzo, J. A. (1967). Inoculation of *Pinus caribaea* with pure cultures of mycorrhizal fungi in Puerto Rico. In *Proc. 14th Int. Union Forest. Res. Organ.* Munich, Vol. 5, pp. 139–48.

Haines, L. W., Maki, T. E., and Sanderford, S. G. (1975). The effect of mechanical site preparation treatments on soil productivity and tree (*Pinus taeda* L. and *P. elliottii* Engelm. var. *elliottii*) growth. *Proc. 4th N. Amer. For. Soils Conf.* (eds. B. Bernier and C. H. Winget) pp. 379–95. Les Presses de l'Universite Laval, Quebec.

Hatch, A. B. (1936). The role of mycorrhizae in afforestation. *J. For.* 34, 22–9.

— (1937). The physical basis of mycotrophy in *Pinus.* Black Rock Forest Bull. No. 6.

Hile, N. and Hennen, J. F. (1969). *In vitro* culture of *Pisolithus tinctorius* mycelium. *Mycologia* 61, 195–8.

Imshenetskii, A. A. (1955). *Mycotrophy in plants.* US Dept. Commer. Transl. TT67-51290 (1967). Washington, DC.

Iyer, J. G., Lipas, E., and Chesters, G. (1971). Correction of mycotrophic deficiencies of tree nursery stock produced on biocide-treated soils. In *Mycorrhizae* (ed. E. Hacskaylo), USDA Forest Serv. Misc. Publ. No. 1189, pp. 233–8.

Kessell, S. L. (1927). The dependence of pine on a biological soil factor. *Emp. For. J.* 6, 70–4.

Krugner, T. L. (1976). Development of ectomycorrhizae, growth, nutrient status, and outplanting performance of loblolly pine seedlings grown in soil infested with *Pisolithus tinctorius* and *Thelephora terrestris* under different fertilization regimes. Ph.D. Thesis, North Carolina State University, Raleigh.

Lamb, R. J. and Richards, B. N. (1971). Effect of mycorrhizal fungi on the growth and nutrient status of slash and radiata pine seedlings. *Aust. For.* 35, 1–7.

— — (1974a). Inoculation of pines with mycorrhizal fungi in natural soils. I. Effect of density and time of application of inoculum and phosphorus amendment on mycorrhizal infection. *Soil Biol. Biochem.* 6, 167–71.

— — (1974b). Inoculation of pines with mycorrhizal fungi in natural soils. II. Effects of density and time of application of inoculum and phosphorus amendment on seedling yield. *Soil Biol. Biochem.* 6, 173–7.

— — (1974c). Survival potential of sexual and asexual spores of ectomycorrhizal fungi. *Trans. Br. mycol. Soc.* 62, 181–91.

Levisohn, I. (1956). Growth stimulation of forest tree seedlings by the activity of free-living mycorrhizal mycelia. *Forestry* 29, 53-9.

— (1958). Effects of mycorrhiza on tree growth. *Soils Fertil.* 21, 73-82.

— (1965). Mycorrhizal investigations. In *Experiments on nutrition problems in forest nurseries* (ed. B. Benzian) For. Comm. Bull. No. 37, Vol. 1, pp. 228-35. HMSO, London.

Lobanow, N. W. (1953). *Mykotrophie der Holzpflanzen.* (Transl. W. Rawald (1960).) Deut. Verlag. Wiss., Berlin. (Original in Russian, Moscow.)

McComb, A. L. (1938). The relations between mycorrhizae and the development and nutrition absorption of pine seedlings in a prairie nursery. *J. For.* 36, 1148-54.

Madu, M. (1967). The biology of ectotrophic mycorrhiza with reference to the growth of pines in Nigeria. *Obeche, J. Tree Club, Univ. Ibadan* 1, 9-16.

Malencon, G. (1938). Les truffes européennes. *Revue Mycol.* 3, 1-92.

Marx, D. H. (1969). The influence of ectotrophic mycorrizal fungi on the resistance of pine roots to pathogenic infections. I. Antagonism of mycorrhizal fungi to root pathogenic fungi and soil bacteria. *Phytopathology* 59, 153-63.

— (1972). Ectomycorrhizae as biological deterrents to pathogenic root infections. *A. Rev. Phytopathol.* 10, 429-54.

— (1973). Growth of ectomycorrhizal and nonmycorrhizal shortleaf pine seedlings in soil with *Phytophthora cinnamomi. Phytopathology* 63, 18-23.

— (1975). Mycorrhizae of exotic trees in the Peruvian Andes and synthesis of ectomycorrhizae on Mexican pines. *Forest Sci.* 21, 353-8.

— (1976). Synthesis of ectomycorrhizae on loblolly pine seedlings with basidiospores of *Pisolithus tinctorius. Forest Sci.* 22, 13-20.

— (1977a). The role of mycorrhizae in forest production. *TAPPI Conf. Papers*, Ann. Mtg., pp. 151-61. Atlanta, Georgia.

— (1977b). Tree host range and world distribution of the ectomycorrhizal fungus *Pisolithus tinctorius. Can. J. Microbiol.* 23, 217-23.

— and Artman, J. D. (1978). Growth and ectomycorrhizal development of loblolly pine seedlings in nursery soil infested with *Pisolithus tinctorius* and *Thelephora terrestris* in Virginia. USDA, Forest Serv. Res. Note SE-256.

— — (1979). *Pisolithus tinctorius* ectomycorrhizae improve survival and growth of pine seedlings on acid coal spoils in Kentucky and Virginia. *Reclam. Rev.* 2, 23-31.

— and Barnett, J. P. (1974). Mycorrhizae and containerized forest tree seedlings. In *Proc. N. Amer. Containerized Forest Tree Seedling Symp.* (eds. R. W. Tinus, W. I. Stein, and W. E. Balmer) pp. 85-92. Great Plains Agric. Council Publ. No. 68.

— and Bryan, W. C. (1969a). *Scleroderma bovista*, an ectotrophic mycorrhizal fungus of pecan. *Phytopathology* 59, 1128-32.

— — (1969b). Studies on ectomycorrhizae of pine in an electronically air-filtered, air-conditioned, plant-growth room. *Can. J. Bot.* 47, 1903-9.

— — (1970). Pure culture synthesis of ectomycorrhizae by *Thelephora terrestris* and *Pisolithus tinctorius* on different conifer hosts. *Can. J. Bot.* 48, 639-43.

— — (1971). Influence of ectomycorrhizae on survival and growth of aseptic seedlings of loblolly pine at high temperature. *Forest Sci.* 17, 37-41.

— — (1975). Growth and ectomycorrhizal development of loblolly pine seedlings in fumigated soil infested with the fungal symbiont *Pisolithus tinctorius. Forest Sci.* 21, 245-54.

— and Daniel, W. J. (1976). Maintaining cultures of ectomycorrhizae and plant pathogenic fungi in sterile water cold storage. *Can. J. Microbiol.* 22, 338-41.

— Bryan, W. C., and Cordell, C. E. (1976). Growth and ectomycorrhizal development of pine seedlings in nursery soils infested with the fungal symbiont *Pisolithus tinctorius*. *Forest Sci.* 22, 91-100.

—— —— (1977). Survival and growth of pine seedlings with *Pisolithus* ectomycorrhizae after two years on reforestation sites in North Carolina and Florida. *Forest Sci.* 23, 363-73.

—— — and Davey, C. B. (1970). Influence of temperature on aseptic synthesis of ectomycorrhizae by *Thelephora terrestris* and *Pisolithus tinctorius* on loblolly pine. *Forest Sci.* 16, 424-31.

—— — and Grand, L. F. (1970). Colonization, isolation, and cultural descriptions of *Thelephora terrestris* and other ectomycorrhizal fungi of shortleaf pine seedlings grown in fumigated soil. *Can. J. Bot.* 48, 207-11.

— Hatch, A. B., and Mendicino, J. F. (1977). High soil fertility decreases sucrose content and susceptibility of loblolly pine roots to ectomycorrhizal infection by *Pisolithus tinctorius*. *Can. J. Bot.* 55, 1569-74.

— Mexal, J. G., and Morris, W. G. (1979). Inoculation of nursery seedbeds with *Pisolithus tinctorius* spores mixed with hydromulch increases ectomycorrhizae and growth of loblolly pines. *Sth. J. appl. For.* 3, 175-8.

— Morris, W. G., and Mexal, J. G. (1978). Growth and ectomycorrhizal development of loblolly pine seedlings in fumigated and nonfumigated soil infested with different fungal symbionts. *Forest Sci.* 24, 193-203.

Mexal, J. and Reid, C. P. P. (1973). The growth of selected mycorrhizal fungi in response to induced water stress. *Can. J. Bot.* 51, 1579-88.

Meyer, F. H. (1964). The role of the fungus *Cenococcum graniforme* (Sow.) Ferd. et Winge in the formation of mor. In *Soil microbiology* (ed. E. A. Jongerius) pp. 23-31. Elsevier, Amsterdam.

— (1973). Distribution of ectomycorrhizae in native and man-made forests. In *Ectomycorrhizae: their ecology and physiology* (eds. G. C. Marks and T. T. Kozlowski) pp. 79-105. Academic Press, New York.

Mikola, P. (1969). Afforestation of treeless areas. *Unasylva* (Suppl.) 23, 35-48.

— (1970). Mycorrhizal inoculation in afforestation. *Int. Rev. For. Res.* 3, 123-96.

— (1973). Application of mycorrhizal symbiosis in forestry practice. In *Ectomycorrhizae: their ecology and physiology* (eds. G. C. Marks and T. T. Kozlowski) pp. 383-411. Academic Press, New York.

Momoh, Z. O. (1973). The problems of mycorrhizal establishment in the Savanna Zone of Nigeria. Res. Paper, Savanna Ser. No. 28. Fed. Dept. For. Res., Nigeria.

— and Gbadegesin, R. A. (1975). Preliminary studies with *Pisolithus tinctorius* as a mycorrhizal fungus of pines in Nigeria. Res. Paper, Savanna Ser. No. 37. Fed. Dept. For. Res., Nigeria.

Moser, M. (1958a). Die Mykorrhiza—Zusammemleben von Pilz und Baum. *Umschau* 9, 267-70.

— (1958b). Die künstliche Mykorrhizaimpfung an Forstpflanzen. I. Erfahrungen bei der Reinkultur von Mykorrhizapilzen. *Forstw. Cbl.* 77, 32-40.

— (1958c). Die künstliche Mykorrhizaimpfung an Forstpflanzen. II. Die Torfstreukultur von Mykorrhizapilzen. *Forstw. Cbl.* 77, 273-8.

— (1958d). Der Einfluss tiefer Temperaturen auf das Wachstum und die Lebenstätigkeit höherer Pilze mit spezieller Berücksichtigung von Mykorrhizapilzen. *Sydowia* 12, 386-99.

— (1959). Die künstliche Mykorrhizaimpfung an Forstpflanzen. III. Die Impfmethodik im Forstgarten. *Forstw. Cbl.* 78, 193-202.

— (1961). Soziologische und ökologische Fragen der Mykorrhiza-Induzierung. *IUFRO Proc. 13th Congress.* Vienna.

— (1963). Die Bedeutung der Mykorrhiza bei Aufforstungen unter besonderer Berücksichtigung von Hochlagen. In *Mykorrhiza* (eds. W. Rawald and H. Lyr) pp. 407-24. Fischer, Jena.

— (1965). Künstliche Mykorrhiza-Impfung und Forstwirtschaft. *Allg. Forstz.* 20, 6-7.

Mullette, K. H. (1976). Studies of Eucalypt mycorrhizas. I. A method of mycorrhizal induction in *Eucalyptus gummifera* (Gaertn. & Hochr.) by *Pisolithus tinctorius* (Pers.) Coker & Couch. *Aust. J. Bot.* 24, 193-200.

Muncie, J. G., Rothwell, F. M., and Kessel, W. G. (1975). Elemental sulfur accumulation in *Pisolithus. Mycopathologia* 55, 95-6.

Otrosina, W. J. (1977). Microbiological and ectomycorrhizal aspects of kaolin spoils. Ph.D. Thesis, School of Forest Resources, University of Georgia.

Palmer, J. G. (1971). Techniques and procedures for culturing ectomycorrhizal fungi. In *Mycorrhiza* (ed. E. Hacskaylo), USDA, Forest Serv. Misc. Publ. No. 1189, pp. 32-46.

Park, J. Y. (1971). Preparation of mycorrhizal grain spawn and its practical feasibility in artificial inoculation. In *Mycorrhiza* (ed. E. Hacskaylo), USDA, Forest Serv. Misc. Publ. No. 1189, pp. 239-40.

Powell, W. M., Hendrix, F. F. Jr, and Marx, D. H. (1968). Chemical control of feeder root necrosis of pecans caused by *Pythium* species and nematodes. *Pl. Dis. Reptr.* 52, 577-8.

Pryor, L. D. (1956). Chlorosis and lack of vigour in seedlings of Renantherous species of *Eucalyptus* caused by lack of mycorrhiza. *Proc. Linn. Soc. N.S.W.* 81, 91-6.

Rosendahl, R. O. and Wilde, S. A. (1942). Occurrence of ectotrophic mycorrhizal fungi in soils of cutover areas and sand dunes. *Bull. ecol. Soc. Am.* 23, 73-4.

Ross, E. W. and Marx, D. H. (1972). Susceptibility of sand pine to *Phytophthora cinnamomi. Phytopathology* 62, 1197-200.

Ruehle, J. L. (1980). Inoculation of containerized loblolly pine seedlings with basidiospores of *Pisolithus tinctorius*. US Dept. Agric. Forest Serv. Res. Note SE-291.

— and Marx, D. H. (1977). Developing ectomycorrhizae on containerized pine seedlings. USDA, Forest Serv. Res. Note SE-242.

Saleh-Rastin, N. (1976). Salt tolerance of the mycorrhizal fungus *Cenococcum graniforme* (Sow.) Ferd. *Eur. J. Forest Pathol.* 6, 184-7.

Schmidt, E. L., Biesbrock, J. A., Bohlool, B. B., and Marx, D. H. (1974). Study of mycorrhizae by means of fluorescent antibody. *Can. J. Microbiol.* 20, 137-9.

Schramm, J. R. (1966). Plant colonization studies on black wastes from anthracite mining in Pennsylvania. *Trans. Am. phil. Soc.* 56.

Schultz, R. C. (1977). Tree growth responses to changes in soil moisture, fertility and microorganisms on difficult sites. *Proc. 5th N. Am. Forest Biol. Workshop*. March, 1978, Gainesville, Florida.

Scott, C. W. (1960). *Pinus radiata*. FAO Forestry and Forest Products Studies (Rome) No. 14.

Shemakhanova, N. M. (1962). *Mycotrophy of woody plants*. US Dept. Commer. Transl. TT66-51073 (1967), Washington, DC.

Singer, R. (1975). *The Agaricales in modern taxonomy*, 3rd edn. Cramer, Vaduz, Germany.

Stevens, R. B. (ed.) (1974). *Mycology guidebook*. University of Washington Press, Seattle.

Takacs, E. A. (1961). Inoculación de especies de pinos con hongos formadores de micorrizas. *Silvicultura* 15, 5-17.

— (1964). Inoculación artificial de pinos de regiones subtropicales con hongos formadores de micorrizas. *Idia, Suplemento Forestal* 12, 41-4.

— (1967). Producción de cultivos puros de hongos micorrizógenos en el Centro Nacional de Investigaciones Agropecuarias, Castelar. *Idia, Suplemento Forestal* 4, 83-7.

Theodorou, C. (1967). Inoculation with pure cultures of mycorrhizal fungi of radiata pine growing in partially sterilized soil. *Aust. For.* 31, 303-9.

— (1971). Introduction of mycorrhizal fungi into soil by spore inoculation of seed. *Aust. For.* 35, 23-6.

— and Bowen, G. D. (1970). Mycorrhizal responses of radiata pine in experiments with different fungi. *Aust. For.* 34, 183-91.

— — (1973). Inoculation of seeds and soil with basidiospores of mycorrhizal fungi. *Soil Biol. Biochem.* 5, 765-71.

Trappe, J. M. (1962). Fungus associates of ectotrophic mycorrhizae. *Bot. Rev.* 28, 538-606.

— (1964). Mycorrhizal hosts and distribution of *Cenococcum graniforme*. *Lloydia* 27, 100-6.

— (1971). Mycorrhiza-forming Ascomycetes. In *Mycorrhiza* (ed. E. Hacskaylo), USDA, Forest Serv. Misc. Publ. No. 1189, pp. 19-37.

— (1977). Selection of fungi for ectomycorrhizal inoculation in nurseries. *A. Rev. Phytopathol.* 15, 203-22.

— and Strand, R. F. (1969). Mycorrhizal deficiency in a Douglas-fir region nursery. *Forest Sci.* 15, 381-9.

van Suchtelen, M. J. (1962). Mykorrhiza bij *Pinus* spp. in de Tropen. *Med. Landbouwhogesch. Opzoekingsstn. Staat Gent.* 27, 1104-6.

Vozzo, J. A. and Hacskaylo, E. (1971). Inoculation of *Pinus caribaea* with ectomycorrhizal fungi in Puerto Rico. *Forest Sci.* 17, 239-45.

Weir, J. R. (1921). *Thelephora terrestris, T. fimbriata*, and *T. caryophyllea* on forest tree seedlings. *Phytopathology* 11, 141-4.

White, D. P. (1941). Prairie soil as a medium for tree growth. *Ecology* 22, 398-407.

Wilde, S. A. (1971). Studies of mycorrhizae in Socialist Republics of Europe. In *Mycorrhiza* (ed. E. Hacskaylo), USDA, Forest Serv. Misc. Publ. No. 1189, pp. 183-6.

Wojahn, K. E. and Iyer, J. G. (1976). Eradicants and mycorrhizae. *Tree Plrs' Notes, Wash.* 27, 12-13.

Worley, J. R. and Hacskaylo, E. (1959). The effect of available soil moisture on the mycorrhizal association of Virginia pine. *Forest Sci.* 5, 267-8.

Zak, B. and Marx, D. H. (1964). Isolation of mycorrhizal fungi from roots of individual slash pines. *Forest Sci.* 10, 214-22.

3 Field performance of *Pisolithus tinctorius* as a mycorrhizal fungus of pines in Nigeria

Z. O. Momoh and R. A. Gbadegesin

Introduction

Earlier reports (Momoh 1972, 1976) have shown that pines cannot grow in Nigeria unless they have been previously infected with suitable mycorrhizal fungi at the nursery stage. Redhead (1974) has traced the early attempts to introduce pines into Nigeria; all failed until soil inocula were also introduced. The three sources of soil inocula were Dola Hill (Zambia), Cameroons, and Oxford (United Kingdom) but it is not absolutely certain which of these sources later led to the successful establishment of pines at Vom on the Jos Plateau in 1954.

The Vom inoculum has subsequently been extensively used in different parts of Nigeria. There has been success with various pines, especially *Pinus oocarpa* and *Pinus caribaea*, in several localities such as the Jos Plateau, the Mambila Plateau, Obudu, Afaka, and Ibadan. These areas are either on fairly high ground (over 250 m above sea level) such as at Afaka, Obudu, and the Jos Plateau or they have a fairly prolonged rainy season such as at Ibadan. In lowlands (less than 200 m above sea level), which also have a rather short rainy season, efforts to grow pines have not been very successful, for example trials are known to have failed in Ejidogari, Mokwa, Lokoja, and Bida. Survival is particularly bad in sandy soils (e.g. Bida) which tend to get hot during the dry season.

One mycorrhizal fungus that commonly fruits in pine plantations in Nigeria is *Rhizopogon luteolus*. Laboratory experiments suggest that this fungus would be killed in the soil after a prolonged dry season coupled with high temperature. In the laboratory, the fungus has an optimum growth at only 23 °C and its mycelium is incapable of growing on Hagem agar at a temperature above 34 °C (Momoh 1972). Soil temperature in the field often approaches or even exceeds this lethal temperature at the peak of the dry season in certain localities in Nigeria (Momoh 1972).

These findings led to the search for a suitable mycorrhizal fungus that would be able to survive the adverse conditions. In the USA Dr D. H. Marx of Forest Sciences Laboratory (Athens, Georgia) had made remarkable progress with *Pisolithus tinctorius*. He had found that the fungus could withstand reasonably high temperatures and was surviving very well in relatively hot areas of Georgia. This fungus was subsequently imported into Nigeria for trial (Momoh and Gbade-gesin 1975). The optimum temperature of the fungus on Hagem agar was found to be about 30 °C and its mycelium was found to grow at 42 °C. Seedlings were later inoculated with this fungus and then planted out in the field. The purpose of this chapter is to indicate the results for nearly three years of field trials.

The layout in the field

Sites were chosen for trials at Miango, Afaka, Bida, Mokwa, and Ejidogari (Fig. 3.1). The first site, Miango, has always been regarded as a good one for pines. It is on a plateau which is over 900 m above sea level. It is cool all the year round (by Nigerian standards) with maximum day temperatures ranging from 27 to 30 °C. Afaka was considered to be a moderately good location for the cultivation of *Pinus oocarpa* and *Pinus caribaea*. It is at an altitude of about 450 m above sea level. It is cool (maximum temperature 21–31 °C) through-out the rainy season (June to October) and during the harmattan period (November to February). It is, however, hot (34–5 °C) during the other months, especially March to May.

Fig. 3.1. Location of the experiments in Nigeria.

Bida, Mokwa, and Ejidogari are warm (30-2 °C) during the rainy season (late May–October). They are hot (34-8 °C) for the rest of the year. All three places are less than 200 m above sea-level.

Seedlings that had been aseptically raised and carefully inoculated with *P. tinctorius* (Momoh and Gbadegesin 1975) were allowed to develop until they mostly attained a height of 10–30 cm and then planted out in the field. As early planting is essential during the rainy season some seedlings of less than 10 cm were used.

At Afaka and Miango 70 seedlings of *Pinus caribaea*, inoculated with *Pisolithus tinctorius*, were planted out on each site, while 40 seedlings were planted at each site in Bida, Mokwa, and Ejidogari. This number was based on the availability of inoculated seedlings. All the seedlings were numbered so that their field performance could be followed. On each site, a similar number of seedlings were planted but inoculated with mycorrhizal soil collected from under *Pinus caribaea* trees at Miango. Owing to shortage of seedlings, the plots set up in 1975 were not replicated.

In 1976 and 1977 a larger number of seedlings (100 per plot) was used. Replications were also possible in 1977. Also, instead of leaving only a 3-m space between *Pisolithus* and soil inoculated plots, a much wider space of 45 m was left between the experimental plot of *Pisolithus* and the control inoculated with mycorrhizal soil collected from Miango. In the 1976 trials, the space left between each experimental block and the control was planted up with four rows of *Eucalyptus camaldulensis*, in the 1977 trials *Eucalyptus torreliana* was used. It was hoped that the wide gap between the experimental blocks and the control would reduce possible cross contaminations.

Assessments and results

The height of each plant was recorded immediately after planting and at regular intervals afterwards. All plots were regularly weeded by hand in accordance with the usual practice in pine plantations in Nigeria.

Records of some of the assessments made in the 1975 plots are shown in Tables 3.1 and 3.2. The first assessment in each case was on the same day of planting in June 1975. The second assessment shown in each table was made in April 1976, after the plants had survived one rainy season and one dry season (a total of ten months) in the field. The third set of assessments were made in February 1978, by which time the plants had survived three rainy seasons in the field. They were also experiencing the third dry season and had had a total of 32 months in the field. Each figure in the table represents the mean for all the plants in a given plot.

Table 3.1. Survival of *Pinus caribaea* (planted in 1975) inoculated with *Pisolithus* compared with those inoculated with Miango soil

Location	Treatment	No. seedlings at planting, June 1975	Survival in April 1976 No.	%	Survival in February 1978 No.	%
Afaka	*Pisolithus* plot	70	67	95.7	64	91.4
	Soil inoculum	70	67	95.7	63	90.0
Miango	*Pisolithus* plot	70	66	94.3	66	94.3
	Soil inoculum	70	60	85.7	59	84.3
Mokwa	*Pisolithus* plot	40	23	57.5	19	47.5
	Soil inoculum	40	14	35.0	3	7.5
Bida	*Pisolithus* plot	40	3	7.5	0	0
	Soil inoculum	40	0	0	0	0
Ejidogari	*Pisolithus* plot	40	22	55.0	19	47.5
	Soil inoculum	40	25	62.5	2	5.0

Table 3.2. Growth of plants inoculated with *Pisolithus* compared with plants inoculated with soil

Location	Treatment	At planting, June 1975. Mean height, cm (a)	April 1976 Mean height, cm (b)	Actual growth, cm (b–a)	February 1978 Mean height, cm (c)	Actual growth, cm (c–a)
Afaka	*Pisolithus* plot	24.0	87.8	63.8	421.9	397.9
	Soil inoculum	26.8	89.8	63.0	400.2	393.4
Miango	*Pisolithus* plot	18	36.6	18.6	228.4	210.4
	Soil inoculum	17.1	34.4	17.3	152.1	135.0
Mokwa	*Pisolithus* plot	7.6	45.9	38.3	317.2	309.6
	Soil inoculum	11.5	41.1	29.6	172.3	160.8
Bida	*Pisolithus* plot	7.9	29.3	21.4	—	—
	Soil inoculum	13.1	—	—	—	—
Ejidogari	*Pisolithus* plot	11.6	41.6	30.0	212.5	200.9
	Soil inoculum	13.8	31.1	17.3	115.0	101.2

The results clearly show that seedlings inoculated with *Pisolithus tinctorius* performed better than those inoculated with soil in all plots. The difference in Miango and Afaka was slight and of no practical value. At Mokwa and Ejidogari *Pisolithus* was dramatically better than the soil inoculum. By February 1978, the survival of plants that had soil inoculum in Mokwa and Ejidogari was only 7.5 per cent and 5.0 per cent respectively, while plants inoculated with *Pisolithus* had a survival of 47.5 per cent in each of the two places.

The result in Bida is of particular interest. All the plants inoculated with Miango soil died before the end of the first dry season in the field. During the same period 7.5 per cent of those inoculated with *Pisolithus* survived. The sandy soil of this area is always hot in the dry season. Even *Pisolithus* appears to be unable to survive for very long.

Although the data in Tables 3.1 and 3.2 are from plots established in 1975, the trend of growth and survival in the plots established in 1976 and 1977 in the various areas are the same. For example, the survival and relative growth of plants as of February 1978 in the plots established at Mokwa and Lafiagi in 1976 are shown in Table 3.3.

Table 3.3. Survival and growth of *Pinus caribaea* (planted in 1976) inoculated with *Pisolithus* compared with those inoculated with Miango soil

Location	Treatment	June 1976		February 1978		Actual growth, cm $b-a$
		No. of seedlings	Mean height, cm a	No. of seedlings	Mean height, cm b	
Mokwa	*Pisolithus*	100	17.9	76	180.1	162.2
	Soil inoculum	100	14.1	24	63.3	49.2
Lafiagi	*Pisolithus*	100	17.0	46	114.1	97.1
	Soil inoculum	100	12.3	20	46.1	33.8

The follow up

After the initial trials in some parts of Northern Nigeria, plots of *Pisolithus tinctorius* have been established in several other locations in Southern Nigeria. The idea is to have banks of *Pisolithus* in different parts of the country where infected soil could be collected for further inoculations. Thus, it is hoped that *Pisolithus tinctorius* will become one of the main mycorrhizal fungi in Nigeria in the near future. At present the following areas scattered all over the country already have trial plots of the fungus: Afaka (near Kaduna), Awi (near Calabar), Bende, Ejidogari (near Jebba), Enugu, Ngwo, Gambari (near Ibadan), Ikom, Lafiagi, Miango (near Jos), Mokwa, and Ore (cf. Fig. 3.1).

A nursery was initiated in Samaru, Zaria, to handle the volume of the pure culture inoculations that had to be done. Ekwebelam is in the process of isolating more mycorrhizal fungi for future trials. His initial efforts in this direction have already been published (Ekwebelam 1977).

Discussion

Pisolithus tinctorius was found to lead to better survival and better growth of *Pinus caribaea* and *Pinus oocarpa* on various sites in Nigeria than other mycorrhizal fungi present in the country. This high efficacy of *Pisolithus* as a symbiont of pines has also been found by Berry and Marx (1976) who discovered that the fungus gave better growth of *Pinus taeda* than other naturally occurring fungi in the southern United States. For this reason efforts are being stepped up to utilize *Pisolithus* more intensively in Nigeria. It is hoped that the mycorrhizal plots being established in various parts of the country would serve as centres for collecting soil inoculum for nursery inoculations in future.

The almost total failure of pine trials in Bida including those inoculated with *Pisolithus tinctorius* shows that more research is needed in this area before pine growth can be recommended there. Although *Pisolithus* survives higher temperatures than a number of other mycorrhizal fungi (Marx and Bryan 1971; Momoh and Gbadegesin 1975) the sandy soil of Bida appears unsuitable even for this fungus. The reason may be partly due to the excessive temperatures of Bida sands in the long dry season and partly due to other reasons such as inadequate organic contents of the soil. Harvey, Larsen, and Jurgensen (1976) found that most (up to 95 per cent) of the active mycorrhizae in Western Montana soil were associated with organic fractions while only 5 per cent occurred in mineral fraction of the soil. Sandy soils are normally poor in organic contents.

It is suggested, therefore, that future studies on this topic should include a more critical analysis of the soil and a simultaneous periodic measurement of the soil temperatures in the experimental areas throughout the year. Khudairi (1969) found that mycorrhizal fungi associated with roots of cultivated and native trees and shrubs, such as date palm, *Zizyhus spina-christi* and *Peganum harmala* in the Mesopotamian desert near Baghdad, appeared to supply the plants with both nutrients and moisture in the dry season—at times taking the place of root hairs. It is therefore feasible that mycorrhizae may have to play such a role for pines in Nigeria during the long dry months experienced in certain parts of the country. If this is the case, survival of the fungus itself will be critical. This, together with the ability of *Pisolithus* to survive relatively high temperatures may be responsible for its better performance than other fungi in areas like Mokwa and Ejidogari. In Miango and Afaka, on the other hand, enough of other mycorrhizal fungi possibly survive the dry season (because of cooler temperatures) to enable them to compete fairly well with *Pisolithus*.

It is further suggested that the efforts to establish *Pisolithus tinctorius* banks in different parts of the country should continue and that the recommendations of Mikola (1975) should form the main basis for mycorrhizal research aimed at solving practical afforestation problems in Nigeria. Also, the recent efforts by Odeyinde and Ekwebelam to try a wide range of fungi should continue. It will be desirable to test any fungi that will be eventually introduced into the field for ability to survive the expected high temperatures in some parts of the country.

Acknowledgements

The bulk of the research was conducted with the research grant provided by International Foundation for Science (IFS). It was this grant that enabled the senior author to visit various parts of the USA in 1974, where he not only saw practical applications of *Pisolithus tinctorius* but also had very valuable discussions with the research workers in this field. We are especially grateful to Dr D. H. Marx of the Forest Science Laboratory, Athens, Georgia, for supplies of inocula and the original suggestion to use *Pisolithus*.

Professor P. Mikola of University of Helsinki, Finland, as IFS co-ordinator for mycorrhizal research, has shown a keen interest in our work and has given valuable advice and encouragement to us at various times.

References

Berry, C. R. and Marx, D. H. (1976). Sewage sludge and *Pisolithus tinctorius* ectomycorrhizae: their effect on growth of pine seedlings. *Forest Sci.* 22, 351–8.

Ekwebelam, S. A. (1977). Isolation of mycorrhizal fungi from roots of Caribbean pine. *Trans. Br. mycol. Soc.* 68, 201–5.

Harvey, A. E., Larsen, M. J., and Jurgensen, M. F. (1976). Distribution of ectomycorrhizae in a mature Douglas-fir/larch forest soil in Western Montana. *Forest Sci.* 22, 393–8.

Khudairi, A. K. (1969). Mycorrhiza in desert soils. *Bioscience* 19, 598–9.

Marx, D. H. and Bryan, W. C. (1971). Influence of ectomycorrhizae on survival and growth of aseptic seedlings of loblolly pine at high temperature. *Forest Sci.* 17, 37–41.

Mikola, P. (1975). Consultancy report on mycorrhiza studies in Nigeria. FAO NIR/73/007.

Momoh, Z. O. (1972). The problem of mycorrhizal establishment in the savanna zone of Nigeria. In: *The development of forest resources in the economic advancement of Nigeria* (eds. C. F. A. Onochie and S. K. Adeyoju) pp. 408–15. Proceedings of Inaugural Conference of Forestry Association of Nigeria, Ibadan 1970.

— (1976). Synthesis of mycorrhiza of *Pinus oocarpa. Ann. appl. Biol.* 82 221-6.

— and Gbadegesin, R. A. (1975). Preliminary studies with *Pisolithus tinctorius* as a mycorrhizal fungus of pines in Nigeria. Research Paper (Savanna series), Federal Department of Forest Research, Nigeria.

Redhead, J. F. (1974). Aspects of the biology of mycorrhizal association occurring on tree species in Nigeria. Ph.D. Thesis, University of Ibadan, Nigeria.

4 Preliminary studies of inoculation of *Pinus* species with ectomycorrhizal fungi in Nigeria

S. A. Ekwebelam

Abstract

Eight mycorrhizal fungi (*Cenococcum graniforme, Lepista nuda, Pisolithus tinctorius, Rhizopogon luteolus, Suillus bovinus, S. granulatus, S. luteus,* and *Thelephora terrestris*) were compared for their effectiveness in stimulating growth of seedlings of *Pinus oocarpa* and *P. caribaea* var. *hondurensis* under semi-controlled and nursery conditions using vegetative mycelia and maize grains as sources of inocula, respectively. The effects of inoculation were assessed after six months' growth.

The results revealed that under semi-controlled conditions, *C. graniforme* and *S. granulatus* stimulated growth of *P. oocarpa* and *P. caribaea* respectively better than the other fungi, whilst in the nursery, *S. bovinus* and *R. luteolus* produced similar effects on both pine species respectively. The improvements in growth brought about by *C. graniforme* and *S. granulatus* under semi-controlled conditions did not manifest themselves under nursery conditions. These results are consistent with the findings of other investigators (see Harley 1969) that inoculation of tree species with ectomycorrhizal fungi can stimulate plant growth, the magnitude of the effects differing between fungi and host species and environmental conditions. The nursery experiment was continued in the field for post-nursery growth studies. To a forester nursery and field inoculation trials are the only really satisfactory means of assessing the efficacy of any nursery management practice. It would, therefore, be necessary to observe how the seedlings perform in the field. This provided the rationale for the field studies which will be reported in due course.

These investigations have been prompted by the desire to find mycorrhizal fungi which in competition with other soil micro-organisms would promote seedling growth in the nursery, and subsequently in the field in problem areas in Nigeria. The Nigeria Forest Service plans to convert large areas of unproductive forest estates into profitable plantations of exotic softwoods using mainly *P. oocarpa* and *P. caribaea*. All plantations so far established indicate that satisfactory growth of these pine species can be achieved, provided a satisfactory mycorrhizal association is developed. Fungi which form mycorrhizae with pines are absent from the soils of the relevant areas, and inoculation of nurseries with soil inoculum is the currently accepted practice. It seems apparent that if mycorrhizal fungi could be found in pure culture able to withstand the climatic conditions prevailing in the relevant areas, and to survive and promote seedling growth in the nursery and field, such fungi would serve as more suitable inocula. Besides, the inocula would be relatively simple to introduce into nursery soil at the time of potting.

The results of the present studies, though exploratory, suggest that the possibility exists of either replacement of, or superimposition on, the existing but less desirable fungi with pure cultures of more appropriate mycorrhizal fungi. Certain species of mycorrhizal fungi have been demonstrated to be more effective than others for these purposes. Inoculation of seeds and seedlings with pure cultures of such fungi would not only result in good plant growth, but also in increased forest productivity. The technical problems of mass production of inoculum for such purposes can be overcome by using the present method or those of other investigators (see Ekwebelam 1975). Björkman (1970) has concluded that if the introduction of certain mycorrhizal fungi could be devised for practical application, the approach would constitute a biological alternative to the use of chemical fertilizers as the stores of mineral nutrients, especially nitrogen, phosphorus, and potassium, which are inaccessible to plants growing in nutrient deficient soils could be readily utilized

References

Björkman, E. (1970). Mycorrhiza and tree nutrition in poor forest soils. *Stud. Forest. Suec.*, Vol. 83.

Ekwebelam, S. A. (1975). Mycorrhizal endophytes of Caribbean pine (*Pinus caribaea* Mor.) and their effects on growth and nutrient status of three host varieties. M.Sc. Thesis, Univ. New England, Armidale, Australia.

Harley, J. L. (1969). *The biology of mycorrhiza*, 2nd edn. Leonard Hill, London.

5 Field performance of *Pinus caribaea* inoculated with pure cultures of four mycorrhizal fungi

A. Ofosu-Asiedu

Introduction

Early attempts to introduce pines to Ghana dating from the late 1940s were unsuccessful. It was not until the mid-sixties when pine seedlings were for the first time inoculated with soil or duff imported from Puerto Rico and British Honduras (Ofosu-Asiedu and Gyimah 1972) that plantings succeeded. Failures of uninoculated pines grown in areas where conifers were not indigenous had been recorded in many places (Mikola 1973).

Soil inoculation of pines has now become a standard practice in Ghanaian nurseries but there are several disadvantages to this method. To ensure that the pines were effectively inoculated with mycorrhizal fungi but at the same time prevent the spread of root pathogens, studies were initiated into the use of pure cultures of mycorrhizal fungi for inoculation. This has been successfully used on a field scale by a number of investigators (Moser 1961; Hacskaylo and Vozzo 1967; Marx and Bryan 1975).

This paper reports the preliminary results of the performance of *Pinus caribaea* Mor. inoculated with four ectomycorrhizal fungi and planted on two sites in Ghana.

Materials and Methods

Fungi

The mycorrhizal fungi used in these experiments were:

Pisolithus tinctorius (Pers.) Coker and Couch, received from Dr Donald Marx, Athens, Georgia, USA.

Thelephora terrestris Ehrh. ex Fr. received from Dr Donald Marx, Athens, Georgia, USA.

Rhizopogon luteolus Fr. received from Dr Donald Marx, Athens, Georgia, USA.

FPG 96 (Ivory Coast fungus), isolated from roots of *P. caribaea* grown in Ivory Coast by the present author.

Preparation of bulk inoculum

The stock cultures for the preparation of bulk inoculum were grown on Hagem medium at 25 °C for three weeks. Mycelium agar discs from stock cultures were used to start mass cultures for inoculation. The mass cultures were raised on peat–vermiculite fortified with Marx–Melin–Norkrans solution with glucose (Marx and Bryan 1975) in 1000 ml Erlenmeyer flasks. The flasks and their contents were autoclaved for thirty minutes and each flask was inoculated with eight mycelium agar discs of one of the four fungi. The flasks were incubated at 25 ± 1 °C for between three and four months at which time the fungi had grown throughout the medium. The procedure for preparing the mass inoculum for soil inoculation was as described by Marx and Bryan (1975).

Processing of seeds

Seeds of *P. caribaea* were sterilized with 30 per cent hydrogen peroxide for 30 seconds, then thoroughly washed with sterile distilled water. Three seeds were sown directly into 2:1 sand–compost medium in 8 X 25 cm polythene bags. Thinning to one seedling per pot was carried out three weeks after germination.

Inoculation of seedlings

A spatula full of the mass inoculum of one of the fungi was used to inoculate each pot when the seedlings were a month old and all the pots receiving that inoculum were placed in one compartment of a greenhouse. Three months after inoculation 12 seedlings were sampled at random from each treatment and examined under a stereo microscope for infection.

Field trials were set up in 1976 and 1977 at two sites, Mesewam near Kumasi and Daboasi near Takoradi. Each trial had four treatments and each treatment was replicated four times with each treatment plot having eighteen seedlings planted to it at a planting distance of 1.8 m. Each plot was separated from the other by a distance of 3.6 m. A completely randomized design was used.

Height and survival of the seedlings were taken at 6- and 12-month intervals from the day the plots were established.

Results

Infection in the nursery was poor for the seedlings inoculated for the 1976 trials. The sampled seedlings gave less than 50 per cent infection in all fungal treatments. At the time of sampling less than half of the seedlings infected with *P. tinctorius* and *R. luteolus* had survived although the seedlings inoculated with FPG 96 and *T. terrestris* had higher survival (Table 5.1). Many of the surviving seedlings were chlorotic.

Table 5.1. Infection and survival of pure culture inoculated seedlings in the nursery four months after inoculation

| | *Pinus caribaea* | |
	No. of seedlings inoculated	Infected and surviving
T. terrestris	250	120
P. tinctorius	250	89
R. luteolus	250	22
FPG 96	250	83

Generally the pure-culture inoculated seedlings in the 1976 trial showed good survival in the first six months in all the two sites (Table 5.2) but the Mesewam trial failed after twelve months while at Daboasi between 55 and 75 per cent of the seedlings survived in all treatments (Table 5.3). The growth in height of these seedlings was extremely variable and ranged from 2.9 to 325 cm. Thus comparative aspects of the experiment was abandoned and the trial at Daboasi was left for observation.

Inoculation was more successful in the seedlings raised for the 1977 trials. All the sampled seedlings were infected. The performance of the seedlings in the nursery after four months was good with most of the plants having lush green needles.

The results for the first six months of the 1977 trial indicate that survival was very high in all treatments and at all the sites (Table 5.4). At Mesewam, seedlings infected with *R. luteolus* and FPG 96 grew faster than those seedlings inoculated with *P. tinctorius* and *T. terrestris*. At Daboasi there were no differences in height between the different fungal treatments. The seedlings grew twice as fast in Daboasi as in Kumasi and differences in height between the two sites started to appear after six months.

Survival one year after transplanting was less at Mesewam than Daboasi and other differences between treatments appeared (Table 5.5). The survival was least for seedlings inoculated with *P. tinctorius* and planted at Mesewam. The seedlings at Daboasi grew better than seedlings at Mesewam. On both sites poor performance was recorded for seedlings inoculated with *T. terrestris*. At Mesewam, seedlings inoculated with FPG 96 grew better than *P. tinctorius* and *R. luteo-*

Table 5.2. Performance of pure culture inoculated pines on two sites after six months of transplanting (1976 trial)

Fungus	Mesewam				Daboasi			
	Seedlings planted	Seedlings surviving (%)	Mean height (cm)	Maximum height (cm)	Seedlings planted	Seedlings surviving (%)	Mean height (cm)	Maximum height (cm)
T. terrestris	36	100	12.5	27.50	36	91.3	16.0	30.0
P. tinctorius	36	97.2	12.9	31.3	36	97.2	15.3	20.5
R. luteolus	27	96.2	12.1	15.2	—	—	—	—
FPG 96	36	91.3	12.2	20.5	36	88.9	13.8	21.3

Table 5.3. Performance of *Pinus caribaea* inoculated with three ecto-mycorrhizal fungi 18 months after planting at Daboasi (1976 trial)

Fungi	Seedlings planted	Seedlings surviving (%)	Minimum height (cm)	Maximum height (cm)	Mean height (cm)
P. tinctorius	36	75	2.9	110.0	25.7
T. terrestris	36	61.1	30.0	325.0	47.6
FPG 96	36	55.6	6.8	240.0	40.8

Table 5.4. Performance of *Pinus caribaea* inoculated with four ecto-mycorrhizal fungi six months after planting on two sites (1977 trial)

	Mesewam		Daboasi	
Fungi	Seedlings surviving (%)	Mean height (cm)	Seedlings surviving (%)	Mean height (cm)
P. tinctorius	99.7	21.9	100	41.0
R. luteolus	100	29.6	100	43.2
FPG 96	100	29.6	100	42.9
T. terrestris	100	24.4	100	38.4

Table 5.5. Performance of *Pinus caribaea* inoculated with four ecto-mycorrhizal fungi 12 months after planting on two sites (1977 trial)

	Mesewam		Daboasi	
Fungi	Seedlings surviving (%)	Mean height (cm)	Seedlings surviving (%)	Mean height (cm)
P. tinctorius	77.8	48.9	100	82.3
R. luteolus	94.4	48.9	100	82.4
FPG 96	93.1	62.1	100	83.7
T. terrestris	94.4	41.7	100	69.7

lus inoculated seedlings. At Daboasi no differences in height were recorded for FPG 96, *P. tinctorius*, and *R. luteolus* inoculated plants.

Discussion

No conclusion can be drawn from the 1976 trials owing to the extreme variability in the Daboasi results and the failure at Mesewam. The variability in the results of the 1976 trial may possibly be attributed to inadequate levels of inoculum used. The complete failure at Mesewam in 1976 was due to a severe drought which occurred soon after the trial was established.

The seedlings in the 1977 trials were less variable in both survival and height and compared to the 1976 trials the seedlings performed better. Site differences in the performance of the seedlings are obvious and the higher rainfall at Daboasi seems to promote more rapid growth.

After a year, however, differences appeared between treatments, FPG 96 inoculated plants grew better than the seedlings treated with the other fungi especially at the Mesewam site. Growth was still greater for the plants grown at Daboasi. *T. terrestris* inoculated seedlings were growing more slowly at all sites although the seedlings inoculated with *T. terrestris* do very well in the greenhouse.

References
Hacskaylo, E. and Vozzo, J. A. (1967). Inoculation of *Pinus caribaea* with pure cultures of mycorrhizal fungi in Puerto Rico. *Proc. 14th Int. Union For. Res. Org. Cong.*, Vol. 5, Section 4, pp. 139–48.
Marx, D. H. and Bryan W. C. (1975). Growth and ectomycorrhizal development of loblolly pine seedlings in fumigated soil infested with fungal symbiont *Pisolithus tinctorius. Forest Sci.* 32, 245–54.
Mikola, P. (1973). Application of mycorrhizal symbiosis in forestry practice. In *Ectomycorrhizae* (eds. G. C. Marks and T. T. Kozlowski) pp. 383–41. Academic Press, New York.
Moser, M. (1961). Soziologische und ökologische Fragen der Mykorriza-Induzierung. *Proc. 13th Int. Union of For. Res. Org. Cong.*, Vienna, Vol. 2, Section 1, pp. 24-6.
Ofosu-Asiedu, A. and Gyimah, A. (1972). Assessment of the presence and types of mycorrhiza in some pine plots in Ghana. *Technical Newsletter*, Vol. 6, Nos. 1 & 2. FPRI, Kumasi.

6 Ectomycorrhiza of *Pinus caribaea* in Uganda

M. A. Chaudhry

Abstract

As in many other tropical countries there is a demand for large quantities of timber for constructional and industrial uses in Uganda. But, by and large, the areas available for raising new plantations are savannahs with restricted rainfall and low fertility. In such areas *Pinus caribaea* promises a considerable success and as such the species has been introduced over a sizeable area over more than a dozen sites.

Plantations on different sites met varying degrees of success. In fact, according to Streets (1962) and Redhead (1974), the trial pine plantations in 1910, 1925, and 1930 failed. The success came in 1946 only when the soil inoculum, obviously carrying mycorrhizal fungi, was brought from Kenya. According to Gibson (1963), the mycorrhizal inoculum was brought to Kenya from South Africa in 1910. At present every pine seedling which is planted out in Uganda forests carries with it, from the nursery, the soil inoculum mixed in the soil which is prepared for raising the plants. Also, quite believably, the dispersal of fungal symbiont must also be taking place through spore dispersal.

In spite of the importance which the pine culture gained in Uganda after 1946, a scientific study of crucially important mycorrhizal association was not initiated. The current study, started in 1977 and expected to conclude in 1980, is thus the first systematic attempt in Uganda to investigate the ontogeny of the ectomycorrhizae on pines.

According to interim results available, yellowish brown ectomycorrhizae exist on *Pinus caribaea* in all parts of Uganda and seem to be well adapted to quite different climatic zones. The vertical distribution varies from just below the undecomposed humus layer to a depth of 45 centimetres. Deep and well-drained soils have greater depths of ectomycorrhizae and the numerical incidence of mycorrhizal roots increases per unit volume of soil in better drained soils. The soils in *Pinus caribaea* plantations are mostly acidic with pH around 5 and at places neutral. In a very few cases they are slightly alkaline. The soils are prevalently deficient in nitrogen. *Suillus granulatus* and *Boletus* sp seem to be most significant fungal symbionts.

References

Gibson, I. A. S. (1963). Eine Mitteilung über die Kiefernmykorrhiza in den Wäldern Kenias. In *Mykorrhiza, Intern. Mykorrhiza-symposium* (eds. W. Rawald and H. Lyr). Fischer, Jena.

Redhead, J. F. (1974). Aspects of the biology of mycorrhizal associations occurring on tree species in Nigeria. Ph.D. Thesis, University of Ibadan, Nigeria.

Streets, R. J. (1962). *Exotic forest trees in the British Commonwealth.* Clarendon Press, Oxford.

7 A review of mycorrhizal inoculation practice in Malawi

R. G. Pawsey

For a period of years prior to 1977, studies on the effects of mycorrhizal inoculation on the growth of seedlings in forest nurseries in Malawi were carried out by silvicultural research staff, and it has been difficult from the documentation available to obtain an accurate impression of the effect of the various treatments tested on different sites at different times. In some experiments, different mycorrhizal treatments were combined with other treatments testing the effect of time of sowing, fertilizer application (time of application, quantity, and type), and watering. The contribution of different mycorrhizal soil mixtures to differences in seedling growth was often obscured by interaction with these other treatments.

The general procedures for the raising of coniferous and eucalypt planting stock in Malawi, are as follows. Slightly raised, level seedbeds are prepared (these being terraced if the nursery site is sloping) and the seed is sown broadcast on the surface of the bed (which consists of 10–15 cm of fine sandy soil). The seed is covered with a layer of fine, sieved, sandy soil, 3 to 4 mm in depth. Conifer seed is usually sown in February–March, and eucalypt seed in September–October. Germination is rapid, normally ten days to two weeks, and the young seedlings are pricked out from the seedbeds and transplanted within a week of emergence into soil contained in polythene tubes. The soil-filled tubes are placed in close-packed narrow terraced blocks (normally fifteen pots wide) which run across the nursery sections. The soil at the bottom of the tubes rests directly on the soil surface of the nursery, which is usually slightly sloped.

During the germination period, the seedbeds are sheltered by thin grass mattresses or by cane screens, and both seedbeds and tubed stock are regularly and adequately watered. The addition of fertilizer to pine nursery stock is usually confined to top dressing with granular NPK formulations, but in most permanent nurseries where sowing is completed in February–March, the addition of fertilizer is normally not considered necessary, unless growth has been unusually slow, when a top dressing may be applied in August. In raising eucalypt planting stock, NPK fertilizer is normally mixed into the soil before tube-filling.

With respect to the general recommendations for mycorrhizal

inoculation of pine tube-stock, the current departmental silvicultural guidebook (Foot 1969) states that:

> Experimental work on raising pine plants with or without mycorrhizal soil has given rather inconsistent results. It is probably not necessary to add inoculated soil where fertile soils are being used or where plants are being raised in permanent nurseries where mycorrhiza is already well established. However, in new nurseries and where low quality sands are being used, mycorrhizal soil is undoubtedly essential. It is recommended as a safety precaution that mycorrhizal soil—collected from an established healthy stand—be added to all basic soils used in raising pines. It is preferable to mix the mycorrhizal (i.e. plantation) soil thoroughly with the bush soil in the ratio of 1:10 before tubes are filled. The practice of 'topping up' with ½–1 inch of mycorrhizal soil is unsatisfactory.

No reference is made in the silvicultural guidebook to the incorporation of appropriate mycorrhizal inoculum into soil used for filling pots prior to pricking out of eucalypt seedlings, and as far as is known, the potting medium for eucalypts is always uninoculated bush soil.

In practice, in pine nurseries, there is considerable variation as to whether or not, or to what degree, mycorrhizal plantation soil is incorporated into the tube mix. In all the large nurseries (both well-established and new) responsible for the supply of planting stock to the 50 000 ha Viphya Pulpwood Project in Northern Malawi, the standard prescription for mixing soil before tube-filling, i.e. 1:10 pine plantation/bush soil, is regularly followed, but in older and smaller nurseries, e.g. in Zomba Mountain (in close juxtaposition to mature pine plantations) the soil for tube-filling is taken entirely from the local plantations. No nursery is known where mycorrhizal inoculation of pine tube-stock soil has been discontinued.

A review of the available accounts of previous mycorrhizal inoculation experiments has confirmed the inconsistent and sometimes contradictory results obtained. A number of factors related to experimental design and to assessment methods make it very difficult to interpret the results obtained. However, the results generally justify the standard recommendations (given above) on the incorporation of plantation soil into the potting medium used for raising pine planting stock.

The difficulties encountered in raising exotic conifers in early trials in Malawi (then Nyasaland) from 1907 onwards, and the later successful introduction of mycorrhizal soil from Rhodesia, have been well described by Clements (1938 and 1941). These details have been summarized more recently in a broader African context by Mikola (1970). It seems certain that soil for inoculation purposes has been introduced into Malawi from very few sources, with the chief (and the only documented) introduction being from a single source in Rhodesia in 1930.

A range of fructifications of fungi which have been recorded as being mycorrhizal in pine and eucalypt crops elsewhere in the world are of common occurrence in plantations of these species in Malawi. Fructifications of *Amanita muscaria*, species of *Boletus*, and other agaric genera are produced regularly in pine plantations during the wet season (November to April), and the fruiting of some of these fungi, notably *A. muscaria*, has apparently increased markedly over the last two or three years. *Thelephora* sp and species of *Scleroderma* have also been observed to be of widespread occurrence in pine plantations. Fructifications of *Scleroderma* spp also occur commonly in eucalypt plantations. Fruiting bodies of what is thought to be *Pisolithus tinctorius* have been collected in eucalypt stands, but this identification has yet to be confirmed.

References

Clements, J. B. (1938). Trials in Nyasaland of introduced pines. *Aust. For.* 3, 21–3.
—— (1941). The introduction of pines into Nyasaland. *Nyasald agric. q. J.* 1, 5–15.
Foot, D. L. (1969). *Silvicultural guidebook.* Department of Forestry, Malawi.
Mikola, P. (1970). Mycorrhizal inoculation in afforestation. *Int. Rev. For. Res.* 3, 123–96.

8 Mycorrhizal inoculation and the afforestation of the Valley of Mexico City

M. Valdés and R. Grada-Yautentzi

Abstract

Mexico needs intensive reforestation activities. Erosion affects in relative proportions 80 per cent of the Mexican territory. Reforestation with pine is expensive because seedlings need forest soil to grow and the cost of moving large volumes of soil is very high. In this soil seedlings will find the mycorrhizal fungi that are a prerequisite to successful establishment in treeless areas. In addition to high costs, this practice is not always successful owing to pathogens introduced within the same soil and the chance that seedlings will not always find their best fungal symbiont.

Pure culture inoculation with selected ectomycorrhizal fungi would be preferable; this practice would reduce the danger of introduction of pathogens with the soil inoculum and would reduce costs of movement of soil. Furthermore, if seedlings start to grow in the eroded soil they would be better adapted when transplanted to the field. At present many of them die after transplantation.

In this study mycelial inoculation of *Pisolithus tinctorius*, *Laccaria laccata*, and the non-mycorrhizal *Lepiota lutea* was conducted at a nursery located in the area of restoration. Fungi were introduced into wooden boxes filled with methyl-bromide fumigated soil mixture. The composition of the mixture was 2:1:1 of local (eroded) soil, sand, and pine bark, plus fertilizers equivalent to 500 kg/ha of 10:10:10 NPK. The species studied were *Pinus michoacana*, *P. montezumae*, *P. pseudostrobus*, and *P. radiata*. After three months, seedlings were transplanted and reinoculated. After three to six months more, ten to twenty seedlings were sampled to make evaluations of growth and ectomycorrhizal infection.

Preliminary results (Table 8.1) showed that healthy pine seedlings can be grown in non-forest soil with pure culture inoculum of ectomycorrhizal fungi. Three months after transplanting, seedlings had a very poor mycorrhizal development. After six months approximately 90 per cent of inoculated seedlings had developed mycorrhizae and about 25 per cent of feeder roots were ectomycorrhizal. It is probable that this percentage will be improved in future experiments. *P. pseudostrobus* was the most promising species. Three months after transplanting seedlings showed only a slight growth increase when

Table 8.1. Percentage growth increase of seedlings of four *Pinus* species inoculated with three fungi growing in a mixture of local (eroded) soil, pine bark, and sand (2:1:1) plus 10:10:10 NPK fertilizer*

Treatment	Measurements	*Pinus*			
		michoacana	*montezumae*	*pseudostrobus*	*radiata*
P. tinctorius	Growth increase % at 3 months	77.4	2.3	–	8.6
	Growth increase % at 6 months	–	–	50.7	12.8
	Ectomycorrhizae formed	bifurcate monopodial (coralloid)	monopodial coralloid	bifurcate coralloid	monopodial bifurcate (coralloid)
L. laccata	Growth increase % at 3 months	63.3	–	13.4	2.1
	Growth increase % at 6 months	–	–	113.7	3.1
	Ectomycorrhizae formed	monopodial bifurcate (coralloid)	monopodial	bifurcate coralloid	monopodial bifurcate (coralloid)
L. lutea	Growth increase % at 3 months	L	6.9	–	12.6
P. tinctorius	Growth increase % at 6 months	L	–	222.2	–
	Ectomycorrhizae formed	–	monopodial	monopodial bifurcate	monopodial bifurcate

*Percentage increase of dry weights of inoculated seedlings in comparison to uninoculated control seedlings. Measurements were from 10 to 20 seedlings.

L = Losses of all seedlings of this treatment due to toppling by incorrect watering.

inoculated with *L. laccata*. After six months inoculated seedlings had greater height and better growth than uninoculated ones. Relative growth increase was 50 per cent when inoculated with *P. tinctorius*, 113.7 with *L. laccata*, and 222.2 with *L. lutea* and *P. tinctorius*.

9 Observations on mycorrhiza of *Pinus merkusii* and its application for afforestation in Indonesia

M. Martini

Abstract

During the last ten years large scale afforestation and reforestation programmes have been carried out in many parts of Indonesia, mainly using *Pinus merkusii.* Pine seedlings for plantations are usually raised in nursery beds and need to be inoculated with pine soil in order to stimulate growth. This raises the problem of how to provide mycorrhizal soil in sufficient quantities. Many other practical problems such as the kinds and types of the inoculum and the possible advantage of using fertilizers still have to be studied to answer important questions relating to the use of *Pinus merkusii* in afforestation. Until recently, however, both the fungi and their possible mycorrhizal association have received little attention. Systematic observations and intensive studies are required to establish a full understanding of the tropical forest mycorrhizae formed by higher Basidiomycetes (Rifai 1975, 1977).

A preliminary study on the relationship of *Pinus merkusii* and mycorrhizal fungi in the nursery has been made (Martini 1972). Two-month-old pine seedlings were grown in natural and in sterilized pine soil. Detailed observations on the root systems showed that the plants grown on non-sterilized medium formed short dichotomous roots with transparent surfaces. It is supposed that these resulted from the presence of mycorrhizal fungi (Melin 1953). Further observation at later ages gave a number of significant differences between the two groups of seedlings, e.g. in the thickness of the stem and the general conditions during the growth period after the age of eight months up to the harvest. A higher content of P_2O_5 and K_2O was found in plants grown on non-sterilized media. Apparently the larger surface area of the roots enabled more nutrient uptake from the soil. The use of pine soil is, therefore, strongly recommended for pine nurseries, for better seedlings and, subsequently, for better growth in the field.

Based on the above observation on mycorrhiza, further research and its application for afforestation has been undertaken, including growing *Pinus merkusii* seedlings in different kinds of soil. In natural broadleaf forest and sterilized soil growth was poor with an average height of 7.6 cm, compared to natural pine soil where growth was successful with an average height of 17.8 cm. Results with *P. merkusii* in soils from *Agathis* and *Shorea* plantations were variable. Further research revealed that well-established seedlings had dichotomous root bundles with yellowish colour, whereas the poor seedlings had dichotomous black roots.

A fungus was isolated from the dichotomous roots of the well-established seedlings. It was cultured on *Quercus soendanensis* sawdust medium and is to be tested for nursery inoculation. Thus, it is possible that pure cultures may be used in the future instead of the mixed soil population from pine plantations.

References

Martini, M. (1972). The relation between the growth of *Pinus merkusii* seedlings and mycorrhiza. (Orig. Indonesian.) For. Res. Inst. Rep. No. 100, Bogor.

Melin, E. (1953). Physiology of mycorrhizal relations in plants. *A. Rev. Plant Physiol.* 4, 325–46.

Rifai, A. M. (1975). Phytopathology and tropical fungi. *Bio. Indonesia* 1, 33–8.

— (1977). Introduction of tropical mycorrhiza forming fungi. *Bio Indonesia* 4, 15–19.

10 Some characteristics of mycorrhizae associated with *Pinus merkusii*

S. Hadi

Introduction

At present about 42 million ha of forest area in Indonesia is devastated and needs to be rehabilitated. Each year the Forest Service plans to replant about 50 000 ha using mainly *Pinus merkusii* Jungh. et de Vriese, an indigenous species of South-East Asia. The production of large numbers of healthy *P. merkusii* seedlings is therefore of great importance.

One of the factors playing an important part in the production of *P. merkusii* seedlings is the establishment of mycorrhizae in the root system. The importance of the mycorrhiza for the good growth of *P. merkusii* seedlings has long been established (Roeloffs 1930; Martini 1972). However, basic information on the characteristics of the mycorrhiza is very sparce. Very little is known about the morphological and anatomical structures, the identity of the fungal symbionts, the isolation and the inoculation techniques, and the effects of soil factors on the development of the mycorrhiza. This paper reviews the available information on the characteristics of mycorrhiza associated with *P. merkusii*.

Morphological and anatomical structures

Mycorrhizae associated with *P. merkusii* seedlings and mature trees are light yellow when young and brownish when mature. In most cases the short mycorrhizal roots are dichotomously branched, but sometimes short branches are simultaneously produced by a short mycorrhizal root. A mantle consisting of several layers of fungal cells develops around the short root. Occasionally, two types of structure may be differentiated in the cortex, i.e. individual dark spherical bodies and chain-like structures. The anatomical characteristics resemble those of ectendomycorrhiza. However, it is not yet established whether the spherical and the chain-like structures are formed by the fungal symbionts. In most cases only fungal mantles and Hartig nets were observed.

Identity of the fungal symbionts

Fruiting bodies of *Boletus* spp are sometimes found in nurseries or in plantations. However, it is not clear yet whether this is the fungus associated with the roots of *P. merkusii*. Attempts to prepare cultures from the fruting body have not yet been successful. No successful results have been obtained when isolation was carried out from the mycorrhizal short roots. Work is still in progress to find appropriate techniques for isolating the fungal symbionts. Successful results would be useful for further studies on the physiological characters of the fungal symbionts, and for selecting the isolates which best promote the growth of *P. merkusii*.

Inoculation techniques

It was once common to transplate 6–8-week-old *P. merkusii* seedlings into transplant beds where trees referred to as 'mother trees' were planted some years earlier. Nine-month-old transplants are usually strong enough to be planted out in the field. This practice has now been abandoned as it is not practical to first transplant the seedlings into polyethylene bags about two weeks or earlier before planting them out into the field. The more common practice in Indonesia today is to transplant 6–8-week-old *P. merkusii* seedlings into poly-ethylene bags (7.5 cm in diameter and 20 cm tall) containing soil collected from *P. merkusii* plantations. However, this can result in root injuries which in turn causes the death of transplanted seedlings, and so another method is now being introduced. Two or three seeds are planted in a polyethylene bag containing soil collected from *P. merkusii* plantations. These are thinned to one seedling when they are old enough to resist damping-off and grown on until ready to be planted out in the field.

Thus the roots of the seedlings are inoculated with the mycorrhizal fungus present in the soil. However, these methods do have some disadvantages. Substantial amounts of soil are continuously removed from the forest for the production of seedlings. These methods are also costly, when an established pine plantation is not available near the nursery. In addition, there is a great risk of introducing soil-borne pathogens. In an attempt to develop an alternative method, Anwar (1977) found that a fairly good mycorrhizal development was obtained when 15 g of natural infested forest soil was used as an inoculum for a seedling transplanted into a polyethylene bag containing 400 g of uninfested soil beneath the root system before the 6-week-old seedling was transplanted. Further work needs to be undertaken to determine the minimum amount of natural infested soil required for successful inoculation.

When four mycorrhizal root cuttings, each 2 cm long, were used as inoculum and planted beneath the root system of a transplanted seedling in a polyethylene bag, Anwar (1977) found the mycorrhiza did not develop. Further study needs to be carried out to determine the appropriate number of root cuttings needed for the successful establishment of the mycorrhiza.

Conditions under which the natural infested soil is stored prior to its use for inoculation may play an important part in the survival of the mycorrhizal fungus. In many cases infested soil may have to be transported over long distances before it is used for inoculation of seedlings in new nurseries. Therefore, study of the effects of the storage period and conditions on the viability of the mycorrhizal fungus would be useful. Similar studies should also be undertaken for mycorrhizal root cuttings to be used for inocula.

Factors affecting the development of mycorrhiza

Results of a preliminary study indicated that when 6-week-old seedlings were transplanted into soil collected from a *P. merkusii* plantation, mycorrhiza started to develop ten weeks after transplantation. Anwar (1977) found that phosphorus did not significantly affect the development of mycorrhiza, but nitrogen stimulated development. The growth of the seedling, measured by its height and its fresh and dry weight, was not significantly affected by the addition of nitrogen and phosphorus to the soil. Studies should also be initiated to determine the effects of adding residues of different plant species to the soil on the development of mycorrhiza.

Conclusion

It is known that the establishment of mycorrhiza on the root system of *P. merkusii* is required for the successful growth of seedlings in the nurseries and that of trees in plantations. However, very little is known about the characteristics of the mycorrhizae associated with this pine species.

Appropriate isolation and inoculation techniques need to be developed to determine the identity of the fungal symbionts and to select the isolates best promoting the growth of *P. merkusii*. It is also desirable to develop an alternative method to replace the current practice of using substantial amounts of soil collected from established pine plantations for inoculum. The latter results in the continuous harmful removal of the soil from the forest and may also cause the introduction of soil-borne pathogens.

References

Anwar, C. (1977). The effects of nitrogen and phosphorus fertilizers and the inoculation method on the development of mycorrhiza and the growth of

Pinus merkusii seedlings. (Orig. Indonesian.) Sarjana Thesis. Faculty of Forestry, Bogor Agricultural University, Bogor.

Martini, M. (1972). Relationship between the growth of *Pinus merkusii* seedlings and mycorrhiza. (Orig. Indonesian.) Forest Research Institute, Bogor, Report No. 100.

Roeloffs, J. W. (1930). About artificial regeneration of *Pinus merkusii* Jungh. et de Vriese and *Pinus khasya* Royle. (Orig. Dutch.) *Tectona* 23, 874–905.

11 The status and future of mycorrhiza research in India

B. K. Bakshi

Our knowledge of the role of mycorrhizal symbiosis in tree nutrition is largely based on research conducted in the temperate forests of the world. The ectomycorrhizal habit is an adaptation in certain species to nutrient deficient soils (Harley 1963). It becomes most effective in stimulating the growth of trees in humus-rich soils which contain low amounts of soluble nutrients. In temperate forests, mycorrhizae occur more frequently in the uppermost horizons, especially in the humus layer, and less so in the basic rich agricultural soils (Harley 1969). Humus-rich soils are highly acidic and therefore favour the optimum development of the acidophilic mycorrhizal fungi. Mycorrhizal fungi are heterotrophic to certain vitamins and amino acids which, however, are made available by soil micro-organisms during decomposition of organic matter (Melin 1953; Hacskaylo 1957).

Temperate forests in India: fir regeneration

The above forest conditions do exist in India, at high altitudes, mainly in the Himalayas. The vegetation is predominantly coniferous, mixed sometimes with hardwoods, principally oaks. In these forests, there is an accumulation of humus, which in mature and overmature fir forests is 25-30 cm or more thick, largely undecomposed, and chiefly responsible for the failure of fir to naturally regenerate. In natural fir forests, seedling recruitment is profuse but the root system is restricted in the humus zone for so long that it is poorly developed, possessing only a tap root and few or no laterals. Up to this stage, mycorrhizal roots are poorly developed or absent. Most seedlings die within 1-3 years owing to low amounts of soluble nutrients in the raw humus zone. The few that survive develop sufficiently long tap roots from the fourth year onwards to reach the mineral layer beneath, where available nutrients stimulate a profuse development of the root system and mycorrhizae in the humus layer. In fir forests the humus layer is known to stimulate mycorrhizal fungi. Non-development of mycorrhiza is probably due to the lack of laterals and short roots. When these eventually develop mycorrhizal associations are encouraged thereby helping seedlings to become established. Ploughing to bring the mineral-rich soil to the top is one of the recommendations suggested for obtaining natural regeneration in fir forests (Bakshi, Reddy, Thapar, and Khan 1972). In view of the difficulties in obtaining natural regeneration in fir, forests are clear-felled and artificially regenerated by raising nursery stock. Seedlings require 5-6 growing seasons to attain a transplantable height of 22 cm; this period can be reduced to 3-4 growing seasons by introducing forest soil into the nursery bed as mycorrhizal inoculum and by foliar application of indoleacetic acid which stimulates root morphogenesis which is a prerequisite to the establishment of mycorrhizal associations (Bakshi 1974).

Tropical forests: soil conditions, nutrient status, and mycorrhiza

In the tropics, forest soils generally contain much less organic material and litter than those of temperate forests. Under favourable conditions of biodegradation, the litter becomes quickly decomposed to release biologically important elements which are released to the soil and eventually made available to plants, thus maintaining the nutrient cycle. The frequency of mycorrhizal roots does not appear to be as common as that found in temperate forests, possibly because in the tropics soil contains little humus and nutrients are present in readily available form minimizing the need for large numbers of mycorrhizal roots. The formation of mycorrhiza in the tropics thus needs investigation as regards the physical and chemical properties of soil, particularly the availability of nutrients.

Natural forests: role of fire and grazing on mycorrhizal fungi

Natural forests are of mixed composition with inherent slow growth and therefore low productivity. In addition all species are not commercially valuable. In the tropics, there is a large proportion of degraded forests which is being increasingly replaced with valuable plantation species. Clear-felling and burning release large quantities of nutrients from the organic matter leaving them in a soluble form which can easily be leached away. Clearing of vegetation also disturbs the natural soil microflora.

The distribution of natural forests and their quality classes are governed principally by climate in which the most important single factor is rainfall. An instance of this is the occurrence of different types of sal forests in northern and central India, the stability of which is governed by various biotic factors including fire and grazing. Teak, another principal species, occurs mainly in central and southern India. Fires are common in most teak forests, and depending on the duration and intensity of the burn, may adversely affect soil microflora including beneficial mycorrhizal fungi. Circumstantial evidence of this was recorded in a young deodar plantation (Bakshi 1957) where the lack or poor development of mycorrhizal roots resulted in stunted growth and paling of needles. This plantation was grown on the site of a burnt coniferous forest. Poor growth was attributed to the disturbance of beneficial soil microflora including mycorrhizal fungi. Grazing, a common occurrence in many forests in India, may cause soil compaction resulting in reduced soil moisture due to water run off and may affect the population of mycorrhizal fungi and mycorrhizal roots. The role of fire and grazing in tropical forests needs to be studied with respect to nutrition through mycorrhizal symbiosis, particularly where exotic pines are raised on the clear-felled and burnt sites of natural forests (see below).

Reforestation: species replacement

Reforestation may be with local species or with exotics. In both cases, mycorrhiza may develop with the indigenous mycorrhizal fungi. Mycorrhizal associations may develop in exotics far removed taxonomically from the species of the indigenous forests. For example, chir pine plantations were successfully raised in Suphkar (Madhya Pradesh), situated more than 1200 miles from its indigenous home in the Himalayas, on former natural sal forests. The pine forests have attained all-India quality I Class and have developed mycorrhiza with fungi forming mycorrhiza in sal.

Subtropical and tropical pines

With changing patterns of demand for industrial woods, particularly for long-fibred pulpwood, exotic tropical and subtropical pines from Central America are under extensive trial in different parts of India including areas outside the Himalayas where no natural pines occur. One such area is the eastern tropical and subtropical region including the plateau of the Eastern Ghats—the dome area—stretching from Orissa in north Andhra Pradesh, parts of Madhya Pradesh (Bastar), and north of Tamil Nadu (Tirumalai Hills). The natural degraded vegetation, comprising sal and other mixed hardwoods, are clear-felled and given over to local inhabitants for intensive shifting cultivation. Repeated cutting, burning, and jhooming cultivation results in the degrading of the site. Higher ground and hill slopes with scanty top soil are planted with pines while the comparatively fertile valley bottoms are left for agriculture. Of the several species of exotic pines raised under trial plantations in the dome area, some species show promise although the failure rate is high. The species mostly develop one type of mycorrhiza, the fungal symbiont conforming to that occurring in short roots of sal and chir pine from Suphkar in which the sal mycorrhizal symbiont is associated. Incidence of mycorrhizal roots was low. This factor as well as others like the degraded site, low availability of soil nutrients, and leaching result in poor stocking rates. Growth of plants is likely to be affected. Mycorrhizal roots also developed poorly in nursery stock.

Mycorrhizal bank

Nursery trials of pines using mycorrhizal soil inoculum from broad-leaved and coniferous forests have shown that, though inoculum from both could successfully induce development of mycorrhizal roots, height and dry weight of pines were significantly greater with soil inoculum from coniferous forests.

This may lead to the conclusion that pines grow better with their own group of mycorrhizal symbionts. Also, mycorrhizal fungi differ in their ability for nutrient uptake. In view of this and the fact that exotic pines have developed mainly one type of mycorrhizal root— the fungal symbiont which belongs to broad-leaved species—a hypothesis was put forward to set up a mycorrhizal nursery bank. In this, the local soil inoculum is fortified with that obtained from beneath subtropical species of *Pinus*, such as *P. roxburghii* (chir pine) and *P. kesiya* (khasya pine) growing in their natural habitat and from as many locations as possible in order to provide a broad spectrum of mycorrhizal fungi in the nursery bed. Soil inoculum containing mycorrhizal fungi from beneath high-altitude Indian conifers may not adapt to subtropical conditions and was not therefore used. Wild seedlings of the two subtropical pines listed above and also tube seedlings of exotic pines with satisfactory development of mycorrhizal roots were planted in the nursery bed at moderate spacing to serve as mother seedlings. The soil inoculum and mother seedlings thus provided a broad spectrum of mycorrhizal fungi. 3-4-week-old seedlings were pricked out in the nursery bed. Field experience has shown that the root system and mycorrhizal roots develop in greater abundance in seedlings raised in nursery beds as compared to tube seedlings. Transplanting seedlings from the nursery bed to tubes may therefore be delayed up to about three months. Tube seedlings constitute the planting stock. While harvesting, an adequate number of seedlings may be left in the bed as a source of inoculum. The nursery soil will possess a rich flora of mycorrhizal fungi to serve as a reservoir for future nursery stock. Nursery management practices on the role of soil pH and moisture on the formation of mycorrhiza, methods of introducing mycorrhiza, control of damping off, role of fungicides on formation of mycorrhiza, and shading of nursery beds to protect seedlings from excess sunlight have already been standardized under Indian conditions. A review of the work on this is given by Bakshi (1974). Mikola (1970) traced the history of mycorrhizal inoculation in greater part of the world including the tropics and laid down guidelines of mycorrhizal inoculation in afforestation. However, little attention has been paid so far to study the characteristics of tropical soils, particularly of degraded sites, in relation to occurrence of mycorrhizal fungi and nutrition through mycorrhiza for success of various reforestation and afforestation projects that are carried out in the tropical countries of the world.

References

Bakshi, B. K. (1957). Forest pathology notes from India: mycorrhiza. *Commonw. phytopath. News.* 3, 43.
— (1974). Mycorrhiza and its role in forestry. PL 480 project report, Dehra Dun.
— Reddy, M. A. R., Thapar, H. S., and Khan, S. N. (1972). Studies on silver fir regeneration. *Indian Forester* 98, 135–44.
Hacskaylo, E. (1957). Mycorrhizae of trees with special emphasis on physiology of ectotrophic types. *Ohio. J. Sci.* 57, 350–7.
Harley, J. L. (1963). Mycorrhiza. *Vistas in botany*, Vol. III, pp. 79–103.
— (1969). *The biology of mycorrhiza*, 2nd edn. Leonard Hill, London.
Melin, E. (1953). Physiology of mycorrhizal relation in plants. *A. Rev. Pl. Physiol.* 4, 325–46.
Mikola, P. (1970). Mycorrhizal inoculation in afforestation. *Int. Rev. For. Res.* 3, 123–96.
Reddy, M. A. R. and Khan, S. N. (1972). Soil amendments and types of inocula on development of mycorrhiza. *Indian Forester* 98, 307–10.

12 The role of mycorrhiza in forestry development in Thailand

C. Khemnark

There is a growing concern in Thailand about the serious degradation in the quality of the forests and the significant reduction in the area of forest land. Concerns for the change in the forest status are expressed by various sectors of the national community and for various reasons. The timber industry is alarmed over a potential shortage of raw material—the familiar 'timber famine syndrome'. Other sectors of the population are concerned about the degradation of watersheds accompanied by water quality problems, loss of wildlife habitat, and deterioration of aesthetic values. The reduction of forest land-base is a natural and not surprising consequence of an increasing requirement for food and space to meet the needs of an expanding human population. The changes which are occurring and which have occurred are not the result of deliberate plan but rather a continuous evolution in response to competitive pressures for use of land and demands for the products of the land.

In the recent FAO Timber Trends Study (1972, completed in mid-1973) it was shown that a total area of 2.7×10^6 ha of forest plantation needs to be established between 1975 and 2000 if the country is not to experience serious deficiencies in the supply of forest products by the end of the century. Until 20 years ago export of timber was one of Thailand's major sources of foreign income. If modern forest management techniques had been introduced then, the potential yield could have been ten times the current production of about 4–5×10^7 cubic metres annually, and timber might have become the largest money-earning commodity of the country. However, this has not happened; the yield of timber is declining, and the forest area is decreasing drastically. The only remedy for this is the establishment of large-scale forest plantations.

Plantation establishment in Thailand started around 1900 with a small-scale teak plantation in the northern provinces. In the early stages the 'taungya system' was practised successfully. Plantations of other tree species started in subsequent years and the area of planting gradually increased. Mycorrhiza was never thought about in plantation practice until 1962 when the Royal Forest Department and the Department of Industry applied for assistance from UNDP for the Raw Material Survey for Pulp and Paper Project; attention was then paid to the establishment of pine plantations. Experiments were made in many parts of the country on planting local and exotic pines and some other fast-growing species including *Eucalyptus.*

There are two indigenous pines in Thailand, *Pinus merkusii* and *Pinus kesiya,* which are found mainly in the northern and north-eastern provinces and occasionally in the central provinces of the country at an elevation of 1000 to about 1800 m above sea level. Associated genera found are *Quercus, Lithocarpus, Castanopsis, Pieris,* and *Eugenia,* and also the deciduous dipterocarps forest. The area of coniferous forest is estimated at 2378 square kilometres (Royal Forest Department 1962).

Experimental planting of pines outside their natural ranges has been carried out in the Suratthani province in the southern peninsular (Bhodhipaksa, Kulchala, and Midewan 1961). It was found that all trees died in the first year of planting, even though inorganic fertilizers were applied after planting. Four exotic pines, namely *Pinus patula, P. caribaea, P. elliottii,* and *P. taeda,* were introduced from Australia by seed. Seedlings of these species were inoculated in the seed bed with mycorrhiza from *Pinus kesiya* mother trees brought as mycorrhizal seedlings from Chiengmai where the pine nursery was established some years ago. When the seedlings were 30 cm high, the seedlings were planted out in the field with success.

Planting pines in their natural ranges was successful in Chiengmai province. The first plantation is now 16 years old, the trees are about 12 m high and 15 cm at breast height. Planting programmes were extending year by year to about 2000 ha per year at present by the Royal Forest Department. The Thai Newsprint Co. is going to start its planting programme in 1980 to prepare raw material for its pulp and paper factories under construction. It is hoped that planting pines outside their natural habitats will be essential and mycorrhiza will play an important role in this large-scale plantation establishment.

Plantations of other species were also started many years ago, but for some species the results seem to be doubtful. It was suspected that mycorrhizae may play a part in these failures, but research has not been undertaken to elucidate the problem.

The present study supported by the International Foundation for Science is attempting to survey the associations between mycorrhizae and pines and some dipterocarps in order to find whether it is possible to inoculate mycorrhizae to make plantation establishment more successful in the future.

From the field investigation and identification of the fruiting bodies growing as symbiont with pines and dipterocarps species, the following mycorrhiza fungi are found: *Pisolithus tinctorius, Russula nigricans, Amanita pantherina, Amanita caesarea, Lactarius volemus,*

Boletus sp, *Coprinus* sp, *Lycoperdon* sp, *Agaricus* sp, and some other species of these genera.

Mycorrhizal roots of four commercial tree species, *Shorea obtusa*, *Dipterocarpus obtusifolius*, *Pinus kesiya*, and *Pinus merkusii*, are isolated and cultured in the laboratory to prepare the inocula for inoculation study and other subsequent experiments.

It is expected that the result of this study will give useful information for pine and dipterocarp plantation establishment outside their natural ranges and also on degraded land, where many new plantations are planned.

References

Bhodhipaksa, P., Kulchala, W., and Midewan, S. (1961). Some observations on mycorrhizal inoculation for pines. Royal Forest Department, Ministry of Agriculture, Bangkok. (In Thai.)

FAO (1972). Present and future forest policy goals: a timber trends study 1970–2000. Report to the Government of Thailand. FAO No. TA 3156. FAO, Rome.

Royal Forest Department (1961). Types of forests of Thailand R. 44. Royal Forest Department, Ministry of Agriculture, Thailand.

13 Ectomycorrhizal fungi of lowland tropical pines in natural forests and exotic plantations

M. H. Ivory

Introduction

All pines form ectomycorrhizal associations with fungi, even in the tropics where most other trees form endomycorrhizal associations (Critchfield and Little 1966; Thomasini 1974). Many studies have been made of pine mycorrhizae in temperate regions, but few have been made in the tropics. This paper reviews the present information on the mycorrhizae of pines in the lowland tropics.

Tropical pines and their environment

Most pines occur naturally in the North Temperate zone or the equable high altitude regions of the subtropical and tropical zones (Critchfield and Little 1966), however, a few species are truly tropical and inhabit the lowland tropics (see below).

Subgenus Pinus *(hard pines)*

Subsect. Sylvestres	*P. kesiya* Royle ex Gord.	S.E. Asia
	P. merkusii Jungh. & de Vriese	S.E. Asia
	P. tropicalis Mor.	Caribbean
Subsect. Australes	*P. cubensis* Grisenb.	Caribbean
	P. occidentalis Sw.	Caribbean
	P. caribaea Mor. v. *caribaea*	Caribbean
	v. *bahamensis*	Caribbean
	v. *hondurensis*	Cent. Amer.
Subsect. Oocarpae	*P. oocarpa* Schiede	Cent. Amer.

Subgenus Strobus *(soft or white pines)*

P. strobus L. v. *chiapensis* Mart.	Cent. Amer.

They usually occur on infertile, well-drained, acidic soils in tropical savannahs with a subhumid/humid climate. Rainfall is low/high (650–4000 mm p.a.) and occurs in the summer. Periods of dry weather often give rise to severe fires (Cooling 1968; Lamb 1973).

Occasionally tropical pines occur on fertile, or poorly-drained, or alkaline soils, or in a perhumid climate. Associated vegetation often includes other ectomycorrhizal woody plants such as *Quercus* spp.

Ectomycorrhizal fungi in natural tropical pine forests

The natural forests of the tropics, including those containing *Pinus* spp, have received little attention from mycologists. Sporadic collections of terrestrial macromycetes have been made, but these have usually been used for taxonomic purposes with little mention being made of their higher plant associates. Such information is mainly documented in taxonomic literature or herbarium records and has not been collated with respect to host species or forest region.

Research is presently underway in Thailand (see Chapter 12), Indonesia (see Chapter 10), and the Philippines (de la Cruz, personal communication) to determine the natural fungal associates of *P. kesiya* and *P. merkusii*. The author is also concerned with a study

Table 13.1. Terrestrial macromycete associates of natural tropical pine forests

---

P. kesiya
  Pisolithus tinctorius (R. de la Cruz in Marx (1977))
P. merkusii
  Boletus sp (Palm 1930), *Suillus* sp aff. *bovinus* (Singer and Morello 1960)
P. oocarpa
  Amanita excelsa, A. flavoconia, A. muscaria, A. ovoidea, A. solitariformis, A. vaginata, 2 Amanita spp, *Boletellus ananas, B. cubensis, Cantharellus* spp, *Coltricia* sp, *Cortinarius* sp, *Gyroporus* sp, *Lactarius* sp, *Phaeogyroporus beniensis, Pisolithus tinctorius*, Polyozellus* sp, *Pulveroboletus hemichrysus, Rhizopogon nigrescens*, 6 *Russula* spp, *Suillus granulatus, S. hirtellus* var. *thermophilus, Tylopilus gracilis*, Tylopilus* sp, *Xerocomus* sp, and 3 undetermined boletes
P. caribaea v. *hondurensis*
  Amanita excelsa, A. flavoconia*, A. gemmata*, A. lilloi*, A. muscaria, A. ovoidea, A. solitariformis, A. verna, 5 *Amanita* spp, *Boletellus cubensis, Boletellus* sp, 3 *Cantharellus* spp, *Collybia peronata, Coltricia* sp, *Geastrum* sp, *Gyroporus castaneus, G. cyanescens, Gyroporus* sp, 2 *Lactarius* spp, *Phaeogyroporus beniensis, Pisolithus tinctorius*, Pulveroboletus hemichrysus, Rhizopogon nigrescens, Russula puiggarii*, 9 *Russula* spp, *Scleroderma* sp*, *S. hirtellus* var. *thermophilus, Suillus* sp aff. *hololeucus*, Thelephora* sp*, *Tylopilus gracilis*, Tylopilus* sp, and 7 undetermined boletes
P. caribaea v. *bahamensis*
  Coltricia sp, *Geastrum* sp, 2 *Hydnum* spp, *Rhizopogon nigrescens*, Scleroderma* sp, *Suillus cothurnatus**, and 3 undetermined boletes

---

*Mycorrhizae formed in aseptic synthesis on *P. caribaea* v. *hondurensis* seedlings also.

of the ectomycorrhizal fungus associates of all tropical pines, but especially those associated with the natural forests of *P. caribaea* and *P. oocarpa* in Central America and the Caribbean. Table 13.1 lists unpublished data from the above study, together with a few published records from elsewhere.

Ectomycorrhizal fungi in tropical pine plantations and nurseries

Species trials with tropical pines have been established in many parts of the tropics, however, only *P. caribaea, P. merkusii,* and *P. oocarpa* have been planted extensively in lowland areas. *P. kesiya* has been planted extensively at somewhat higher elevations.

Many general reports have been published concerning mycorrhizae of pines in tropical countries, but few give details of the host species, mycorrhiza morphology, or associated fungi. Tables 13.2, 13.3, 13.4, and 13.5 have been compiled from the few detailed reports available and from unpublished observations of the author.

Mycorrhizae have additionally been aseptically synthesized on (a) *P. caribaea* with cultures of *Rhizopogon roseolus* ex Sweden, *Corticium bicolor* ex Finland, and *Suillus cothurnatus* ex USA in Puerto Rico (Hacskaylo and Vozzo 1967), and *Pisolithus tinctorius* in the USA (Marx 1977), and on (b) *P. oocarpa* with cultures of *Amanita rubescens, Suillus granulatus, S. luteus, Rhizopogon luteolus, R. roseolus, Russula emetica,* and *Cenococcum geophilum* in Costa Rica (Volkart 1964).

Some early reports of mycorrhizae of *P. caribaea* probably relate to *P. elliottii* Engelm. (Mukherji and Thapar 1961; Young 1937), and others concerning the mycorrhizae on *P. cubensis* in Guatemala (Palm 1930) and *P. occidentalis* in Honduras (Marx 1977) are clearly erroneous.

Origin of ectomycorrhizal fungi in exotic plantations and nurseries

Ectomycorrhizal fungi occur on about 5 per cent of tropical tree species (Redhead 1968; Thomazini 1974), especially in the families Caesalpiniaceae, Dipterocarpaceae, and Myrtaceae. In a few instances these indigenous fungi can infect introduced pines (Bowen 1963), but not in the majority of cases (Redhead 1974). In most areas, therefore, indigenous fungi are not likely to form a significant source of ectomycorrhizal inoculum for introduced pines.

Exotic fungi were probably introduced, along with their native hosts, on ornamental seedlings by early settlers long before the importance of mycorrhizae was realized (see Chapter 1). Deficiencies were not encountered until attempts were made to grow pines in new

Table 13.2. Mycorrhizae and associated fungi in exotic *P. caribaea* nurseries (a) and plantations (b)

Country	Mycorrhiza morphology	Associated fungi
Malaysia	(a) Mycorrhizae white/coral/pale brown, bifurcate/coralloid. Rhizomorphs white, ± numerous (Ivory 1975; Hong 1976)	*Rhizopogon villosus* (Ivory 1975; Hong 1976)
	(b) Mycorrhizae white, simple, very long, in terminal clusters. Mantle smooth, 10–12 μm thick. Hartig net well developed (Griffiths 1965)	*Coltricia cinnamomea, Inocybe* sp, and *R. villosus* (Ivory 1975)
Sri Lanka	Mycorrhizae rusty-brown, coralloid. Mantle and Hartig net well developed. Rhizomorphs numerous (Muttiah 1972)	*Rhizopogon* sp (Mikola in Redhead 1968) 1 isolate ex mycorrhizae (Muttiah 1972)
Australia	—	(b) *R. roseolus, S. granulatus* (Bevege in Ekwebelam 1977) 10 isolates ex mycorrhizae (Ekwebelam 1977) *Boletus* sp, *Coltricia* sp, *P. tinctorius, Tylopilus* sp
New Zealand	—	*R. roseolus* (Rawlings 1951)
Zaire	(b) Dominik types A, F, and H (Thoen 1974)	*S. granulatus, Tylopilus* sp (Thoen 1974)
Nigeria	(a) Mycorrhizae white. Mantle loose. Rhizomorphs numerous (Chapter 1)	(b) *Rhizopogon* sp (Mikola 1970)
Ghana	(b) Mycorrhizae dark-brown, bifurcate/loose coralloid	*P. tinctorius*
Kenya	(a) Mycorrhizae white or pale red-brown	—
Uganda	—	(b) *Amanita* sp, *Boletus* sp, *S. granulatus* (Chapter 6)
Trinidad	(a) Mycorrhizae pale-brown. Mantle compact, 20–40 μm thick. Hartig net well developed (Mikola 1970)	*Rhizopogon* sp (Lackham 1972)

Table 13.2 (*cont.*)

Country	Mycorrhiza morphology	Associated fungi
Puerto Rico	(a) Mycorrhizae greyish-white, bifurcate. Mantle thin	(a) *Thelephora terrestris* 2 isolates ex mycorrhizae (Vozzo and Hacskaylo 1971) (b) *T. terrestris* (Vozzo and Hacskaylo 1971), *Geastrum* sp, *P. tinctorius, Rhizopogon nigrescens, Russula brevipes, Scleroderma citrinum*, and *S. geaster*
Brazil	—	*P. tinctorius* (Hodges in Marx 1977)

Table 13.3. Mycorrhizae and associated fungi in exotic *P. oocarpa* nurseries (a) and plantations (b)

Country	Mycorrhiza morphology	Associated fungi
Malaysia	(a) White, coralloid (Ivory 1975)	(b) *S. granulatus, Inocybe* sp (Ivory 1975)
Australia	—	(b) *Rhizopogon* sp, *P. tinctorius*
Zaire	(b) Dominik types B & F (Thoen 1974)	—
Nigeria	—	(a) *P. tinctorius* (Momoh and Gbadegesin 1975) *R.* (*luteolus*?) (Momoh 1976) (b) *Rhizopogon* (*luteolus*?) (Momoh 1976) 2 *Rhizopogon* spp
Zambia	—	(a) *T. terrestris*
Peru	(b) 6 types (Marx 1975)	—

Table 13.4. Mycorrhizae and associated fungi in exotic *P. kesiya* nurseries (a) and plantations (b)

Country	Mycorrhiza morphology	Associated fungi
Malaysia	(a) White, coralloid, squat (Hong 1976) (b) White, bifurcate/loose coralloid, squat. Mantle 10–12 μm thick; Hartig net well developed (Griffiths 1965)	— (b) *S. granulatus, Boletus pernanus* (Ivory 1975)
Zaire	(b) Dominik types B, F, & H (Thoen 1974)	*S. granulatus, Porphyrellus* sp (Thoen 1974)
Zambia	(a) Ectendomycorrhizae with almost no mantle, and a coarse Hartig net with heavy intracellular penetration (Chapter 1) —	(a) *T. terrestris* (b) *R. luteolus* (CMI unpublished) *Inocybe* sp, *S. granulatus*

Table 13.5. Mycorrhizae and associated fungi in exotic *P. merkusii* nurseries (a) and plantations (b)

Country	Mycorrhiza morphology	Associated fungi
Malaysia	(a) White, short, unbranched (Ivory 1975) (b) White, bifurcate/loose coralloid. Mantle 25–30 μm, two-layered (Griffiths 1965)	(b) *Amanita sychnopyramis, S. granulatus* (Ivory 1975)
Indonesia	Pale yellow/brown, bifurcate/coralloid (Hadi 1978)	*Boletus (pallidus?)* Frost (Palm 1930; Chapter 10)

areas on a larger scale from seed (see Chapter 1). These deficiencies were usually overcome by the introduction of fresh pine soil, or duff, into nursery beds, or by the use of 'mother seedlings'. The history of these importations is well documented for some countries (Lackham 1972; Mikola 1970; Vozzo and Hacskaylo 1971), but not for others. The numbers and sources of these importations, and the way in which they were handled, would obviously have had a great influence on which fungi were successfully introduced to any given area. In many cases the fungi have been able to adapt successfully to their new hosts and environments, however a few instances of their lack of adaptability to high temperatures have been reported (Marx 1977; Muttiah 1972; see also Chapter 3).

Introductions of selected fungi, as basidiocarps, spores, or isolates in culture, have also been made (Hacskaylo and Vozzo 1967). The advantages and disadvantages of these methods have recently been reviewed by Trappe (1977). These sources of inoculum obviously have merit where the establishment of particular fungi is required to overcome adverse site conditions. It is also far more desirable from plant quarantine considerations.

Acknowledgements

I thank the Forestry Departments of the Bahamas, Belize, Guatemala, Honduras, and Nicaragua for permission to survey their forests, and for the facilities they generously provided. I also extend my gratitude to Drs D. N. Pegler and J. M. Trappe for the determination of my collections.

References

Bowen, G. D. (1963). The natural occurrence of mycorrhiza fungi for *Pinus radiata* in South Australian soils. CSIRO Division of Soils, Report No. 6.

Cooling, E. N. G. (1968). *Pinus merkusii.* Fast-growing timber trees of the lowland tropics No. 4. Commonwealth Forestry Institute, Oxford.

Critchfield, W. B. and Little, E. L. (1966). Geographic distribution of the pines of the world. USDA For. Serv., Misc. publ. 991.

Ekwebelam, S. A. (1977). Isolation of mycorrhizal fungi from roots of Caribbean pine. *Trans. Br. mycol. Soc.* 68, 201–5.

Griffiths, D. A. (1965). The mycorrhiza of some conifers grown in Malaya. *Malay Forester* 28, 118–21.

Hacskaylo, E. and Vozzo, J. A. (1967). Inoculation of *Pinus caribaea* with pure cultures of mycorrhizal fungi in Puerto Rico. Proc. 14th Congr. IUFRO, Munich, Vol. 5, pp. 139–48.

Hong, L. T. (1976). Mycorrhizal short-root development on *Pinus caribaea* seedlings after fungicidal treatment. *Malay Forester* 39, 147–56.

Ivory, M. H. (1975). Mycorrhizal studies on exotic conifers in West Malaysia. *Malay Forester* 38, 149–52.

Lackham, N. P. (1972). *Pinus caribaea* in Trinidad and Tobago. Report For. Div., Trinidad.

Lamb, A. F. A. (1973). *Pinus caribaea.* Fast-growing timber trees of the lowland tropics No. 6. Commonwealth Forestry Institute, Oxford.

Marx, D. H. (1975). Mycorrhizae of exotic trees in the Peruvian Andes and synthesis of ectomycorrhizae on Mexican pines. *Forest Sci.* 21, 353–8.

—— (1977). Tree host range and world distribution of the ectomycorrhizal fungus *Pisolithus tinctorius. Can. J. Microbiol.* 23, 217–23.

Mikola, P. (1970). Mycorrhizal inoculation in afforestation. *Int. Rev. For. Res.* 3, 123–96.

Momoh, Z. O. (1976). Synthesis of mycorrhiza on *Pinus oocarpa*. *Ann. appl. Biol.* 82, 221-6.

— and Gbadegesin, R. A. (1975). Preliminary studies with *Pisolithus tinctorius* as a mycorrhizal fungus of pines in Nigeria. *Res. pap.* (Savanna ser.) No. 37, Fed. Dept. For. Res., Nigeria.

Mukherji, S. K. and Thapar, H. S. (1961). Mycorrhizae in seven exotic conifers growing in New Forest. *Indian Forester* 87, 484-8.

Muttiah, S. (1972). Initial observations on the introduction of *Pinus caribaea* in Ceylon and certain rooting, transpiration and mycorrhizal studies on seedlings of two provenances of this. *Ceylon Forester* 9, 98-141.

Palm, B. (1930). *Pinus* and *Boletus* in the tropics. *Svensk bot. Tidskr.* 24, 519-23.

Rawlings, G. B. (1951), The mycorrhizas of trees in New Zealand forests. *Forest Res. Notes, Wellington* 1, 15-17.

Redhead, J. F. (1968). Mycorrhizal associations in some Nigerian forest trees. *Trans. Br. mycol. Soc.* 51, 377-87.

— (1974). Aspects of the biology of mycorrhizal associations occurring on tree species in Nigeria. Ph.D. thesis, University of Ibadan, Nigeria.

Singer, R. and Morello, J. H. (1960). Ectotrophic forest tree mycorrhizae and forest communities. *Ecology* 41, 549-51.

Thoen, D. (1974). Premières indications sur les mycorrhizes et les champignons mycorrhiziques des plantations d'exotiques du Haut-Shaba (République du Zaïre). *Bull. Rech. Agron., Gembloux* 9, 215-27.

Thomazini, L. I. (1974). Mycorrhiza in plants of the 'Cerrado'. *Pl. Soil* 41, 707-11.

Trappe, J. M. (1977). Selection of fungi for ectomycorrhizal inoculations in nurseries. *A. Rev. Phytopathol.* 15, 203-22.

Volkart, C. M. (1964). Formation of mycorrhizae on Central American pines under controlled conditions. *Turrialba* 14, 203-5.

Vozzo, J. A. and Hacskaylo, E. (1971). Inoculation of *Pinus caribaea* with ectomycorrhizal fungi in Puerto Rico. *Forest Sci.* 17, 239-45.

Young, H. E. (1937). *Rhizopogon luteolus*, a mycorrhizal fungus of *Pinus*. *Forestry* 11, 30-1.

14 An observation on the occurrence of pine mycorrhiza-like structures in association with bracken rhizomes

S. Kondas

The bracken *Pteridium aquilinum* (L.) Kuhn is an invasive rhizomatous fern of the open, moist plateau grasslands of the Western Ghats in South India. In young plantations of *Pinus patula* Schiede & Deppe raised in grasslands, pine laterals are sometimes noticed tunnelling through fragments of decayed bracken rhizome. In one particular instance many branches arose from a portion of a lateral covered by rhizome to a length of 15 cm. This, together with the occurrence of coralloid mycorrhiza-like structures, contrasted sharply with the very few branches and extremely few mycorrhiza-like forked roots on the uncovered region of the same root (Fig. 14.1).

Of the many soluble and insoluble carbon sources of ectomycorrhizal fungal metabolites studied by Palmer and Hacskaylo (1970),

Fig. 14.1. Coralloid mycorrhiza-like structures on a pine root within the bracken rhizome casing.

the most important ones were found to be D-glucose and D-mannose. Bracken rhizome contains both, the former comprising 60 per cent of the nucleotide sugars and the latter 5 per cent, besides a number of other sugars and sugar derivatives which may very well form energy and nutrient sources for the mycorrhizal fungi.

Ectomycorrhizal fungi provide their hosts with hormones (auxins, cytokinins, gibberellins) and growth regulating B-vitamins. The excess hormonal substances promote root morphogenesis. Slankis (1958), using indoleacetic acid of different concentrations, showed varying degrees of morphogenesis on attached roots of *Pinus strobus*. At lower concentrations dichotomy of roots occurred, while at five to ten times higher concentration structures closely resembling simple and coralloid mycorrhizae were formed. Turner (1962) in his study of 53 fungal species found that different fungal exudates differed in their effect on excised roots of *Pinus sylvestris*. Exudates of *Amanita muscaria* stimulated elongation of excised radicles and induced lateral root formation; *Suillus* (*Boletus*) *variegatus* caused prolific induction of root initials. Slankis (1951) and Palmer (1954) reported that exogenous auxin stimulated elongation of pine roots at low concentration, inhibited it at higher concentration, and at still higher concentration induced new laterals in increasing numbers.

The fleshy bracken rhizome traversed by a network of vascular strands (meristeles) and bands of sclerenchyma holds a rich store of nucleotide sugars, carbohydrates, and a number of other compounds which provide a good energy source for invading mycorrhizal fungi. The fibrous rhizome 'casing' facilitates accumulation of exudates (hormones and growth regulators) of host as well as of fungus and probably favours early effective symbiosis. Since the concentration of hormonal substances remains low in the early stages of infection, it is likely that stimulation of both long and short roots from the lateral is possible. As the concentration builds up within the confines of the rhizome, further development of long roots is inhibited and the frequency of short roots increases. As the 'casing' proves an effective barrier against loss of exudates, structures closely resembling coralloid mycorrhizae develop in large numbers at levels of increased concentration.

The fungal component of the mycorrhiza-like structures was not identified. However, circumstantial as the evidence may be, the possibility of a true symbiosis is strong. The premise for this is the occurrence of fruiting bodies of mycorrhizal fungi like *Amanita muscaria*, *Suillus* spp, *Rhizopogon* spp, *Thelephora terrestris*, *Russula* spp, etc. which are strictly confined to pine plantations and not found in the adjacent wattle and *Eucalyptus* plantations in this locality.

Obviously this is an oversimplified account of the occurrence of mycorrhiza-like structures and the synergistic and antagonistic influences of the partners in causing the biochemical and biological transformations. The phenomenon is probably the result of a multitude of factors which take place within the confines of the decayed rhizome.

As the rhizome appears to be a rich source of nutrients and other compounds, and a good medium for fungal growth and development it may be possible to use it as a growth promoter in the nursery, either in mother bed or in polycontainers.

References

Duncan, H. J. and Jarvis, M. C. (1976). Nucleotides and related compounds in bracken. Taxonomy and phytogeography of bracken—a review. *J. Linn. Soc. (Botany)* 1-34.

Palmer, J. G. (1954). Mycorrhizal development in *Pinus virginiana* as influenced by growth regulators. Ph.D. Thesis, George Washington University, Washington, DC.

— and Hacskaylo, E. (1970). Ectomycorrhizal fungi in pure culture. I. Growth on single carbon sources. *Physiologia Pl.* 23, 1187-97.

Slankis, V. (1951). Über den Einfluss von B-Indolylessigsäure und anderen Wuchsstoffen auf das Wachstum von Kiefernwurzeln. I. *Symb. bot. upsal.* 11, 1.

— (1958). The role of auxin and other exudates in mycorrhizal symbiosis of forest trees. In *Physiology of forest trees* (ed. K. V. Thimann) pp. 427-43. Ronald Press, New York.

Turner, P. D. (1962). Morphological influences of exudates of mycorrhizal and nonmycorrhizal fungi on excised root cultures of *Pinus sylvestris* L. *Nature, Lond.* 194, 551.

15 The use of chemicals for weed control in forest tree nurseries

C. Iloba

Abstract

Ectomycorrhizal symbiosis is now accepted as an important tool in forest management all over the world. Its potency in the tropics is highlighted for two reasons: (1) The economic situation of Third World countries and the pressing need to meet limited forest produce requirements, requires the establishment of forest plantations of exotic, fast-growing, obligatory-ectomycorrhizal tree species. (2) As few tropical forest trees are ectomycorrhizal the possibility for natural inoculation, as occurs in the predominantly ectomycorrhizal forest zones, is slight.

The establishment of ectomycorrhizae in the tropics, therefore, requires artificial silvicultural manipulations. The forest tree nurseries determine, to a large extent, the success or otherwise of any afforestation programme, as they should supply high quality planting stock with well developed ectomycorrhizae at the time of lifting.

In recent times pathological and economic considerations have made the use of various pesticides and herbicides in forest nurseries imperative. This is done with little or no consideration for the economically significant, non-target micro-organisms such as the ectomycorrhizal fungi. The labour situation in Nigeria since the payment of the 'Udoji award' has made manual weeding too expensive. Herbicide application is now regarded as a labour-saving and economic alternative. The effects of biocides on mycorrhizae differ. In the majority of the cases herbicides exhibit inhibiting and fungitoxic effects on mycelial production of ectomycorrhizal fungi as well as to the development of essential mycotrophic organs in young seedlings. This paper, summarizing studies by the author (Iloba 1971, 1975, 1976, 1977a,b), examines the possibilities of tolerance to some herbicides by older seedlings, based on the parameter of ectomycorrhizal development.

The sodium salt of chlorinated phenoxy-acetic acid (2,4D) at concentrations of 0.5 and 1 g/m^2, and trifluralin at rates of 0.2 and 0.4 ml/m^2 were used in the investigations. Pots filled with infected humus were treated with the herbicides at different concentrations. The test plants were well developed and highly ectomycorrhizal one-year-old seedlings of *Pinus sylvestris* L. and three-year-old *Picea abies*

(L.) Karst. *P. sylvestris* were harvested after 10 months in the green-house and *P. abies* was left in the field and harvested at five different periods. All the seedlings were subjected to thorough morphological examination and compared with controls.

The results showed that, age and species differences notwithstanding, the treated seedlings were affected by the different concentrations of herbicides. 2,4D caused abnormal elongation of the lateral roots and root hairs at higher concentrations. This was not the case with the untreated seedlings, whose laterals remained well developed and highly mycorrhizal. Trifluralin adversely affected the development of ectomycorrhizae. Even the established infected roots were shrunken, deformed, and killed. Newly initiated short roots were uninfected at the time of seedlings' evaluation. In general, the lateral roots in treated soils differed morphologically from controls.

While the absence of the marked heterorhizy for the differentiation of short and long roots in treated seedlings was evident, the negative effect of the herbicides was accentuated by the total suppression of dichotomous branching in *P. sylvestris*. The deleterious effects of the two herbicides come as no surprise because their phytotoxicity, particularly on weeds, is familiar. The sensitivity of most ectomycorrhizal fungi to these herbicides, as was observed earlier, is confirmation of their fungitoxic effects. So far ectomycorrhizal fungal symbionts are good indicators for forecasting which herbicides will show harmful effects to the physiologically important symbiosis in the seedlings. Perfect control of economically important weeds in the nurseries is an important silvicultural objective and this makes the application of efficaceous herbicides attractive and imperative.

The production of seemingly well-developed but non-ectomycorrhizal seedlings negates the ultimate objective of nursery management, i.e. the production of high quality, viable planting stock. If herbicides are to be used in intensified silvicultural management, the trauma for the seedlings should be reduced by avoiding the use of highly toxic compounds that heighten the attendant risks.

Though the investigations were carried out in a temperate zone with temperate soils and plants the results obtained are very pertinent to practical silviculture in the tropics. Both plant production in general and large scale forest nursery stock production are using chemical herbicides. In view of the urgency to satisfy the various needs for forest products, Third World countries can scarcely afford the luxury of the attendant risks associated with highly efficaceous herbicides. Biocides intended for use in forest tree nurseries should primarily be selected on the basis of their potential effect on ectomycorrhizae.

References

Iloba, C. (1971). Einflüsse der Pflanzenschutzmittel auf die Mykorrhizabildung bei Waldbäumen. Diss. Sekt. Forstw., Techn. Univ. Dresden.

—— (1975). Aspects of herbicidal effects on the ecosymbiotic microorganisms of the forest plants. *Eur. J. Forest Pathol.* 5, 339-43.

—— (1976). The effects of some herbicides on the development of ectotrophic mycorrhiza of *Pinus sylvestris* L. *Eur. J. Forest Pathol.* 6, 312-18.

—— (1977a). The effect of trifluralin on the formation of ectotrophic mycorrhizae on some pine species. I. Toxicity to mycorrhiza-forming fungi. *Eur. J. Forest Pathol.* 7, 47-51.

—— (1977b). The effect of trifluralin on the formation of ectotrophic mycorrhizae in some pine species. II. On the toxicity to ectotrophic mycorrhizae. *Eur. J. Forest Pathol.* 7, 172-7.

PART III

Mycorrhiza in natural vegetation

16 Mycorrhiza in natural tropical forests

J. F. Redhead

Introduction

Interest in the mycorrhizal associations of tropical plants began almost one hundred years ago, when Treub (1885) recorded the vesicular-arbuscular mycorrhizal association on sugar cane in Java. In the same year Frank (1885) coined the term 'mycorrhiza' and realized that the association of a fungus and a tree root was a naturally occurring, non-pathogenic association.

The first extensive survey of the occurrence of the mycorrhizal association in tropical plants was carried out in 1896 by Janse in Java. He studied Bryophytes, vascular Cryptogams, Gymnosperms, Monocotyledons, and 38 species of woody Dicotyledons. He found that 69 of the 75 species examined, including all the woody Dicotyledons, had characteristic endotrophic mycorrhizal associations. Janse also studied the morphology of the endophyte and gave the names 'vesicles' and 'sporangioles' to the special organs which he observed. Vesicles were small, round or elongated bladder-like structures formed terminally on hyphae within the root cortex; sporangioles were clumps of fungal material, cauliflower-like in appearance, which formed within the root cells at some distance from the root tip. Janse illustrated his account with careful drawings which leave no doubt he was describing the vesicular-arbuscular type of mycorrhizal association.

Despite this promising start, further interest was sporadic and the next extensive survey of mycorrhizal associations in the tropics was carried out fifty years later by Johnston (1949). He examined 93 species, including 13 species of forest trees, and observed that 80, including all the forest trees, had endotrophic mycorrhizal associations.

The intervening period had been a time of considerable speculation on the symbiotic relationship, based largely on descriptive work, and some of the contemporary leading pathologists (Hartig 1888; Wakker and Went 1898) were convinced that the fungus was parasitic. It was only gradually that experimental evidence accumulated indicating that the mycorrhizal association had a profound, often beneficial effect on plant growth, and under certain conditions might be essential for plant growth (Gerdemann 1968; Harley 1968, 1969; Mosse 1963).

Richards and other tropical rain forest ecologists, familiar with Janse's study, saw great significance in the fact that a large proportion of rain-forest plants contained mycorrhizal fungi. They realized that rain-forest soils were generally extremely poor in plant nutrients and that the apparent richness of plant life in these forests was due to the tightly closed cycle of nutrients. Richards (1952) wrote, 'There is little reason to doubt that a careful examination of the root system of rain-forest plants would show that a large percentage are associated with mycorrhizal fungi.' Such comments have stimulated further research into the occurrence of mycorrhizal associations in tropical plants.

This paper attempts to summarize the work on naturally occurring mycorrhizal associations in the tropics, carried out largely during the past 20 years. It does not attempt to review the studies on orchid mycorrhizae as this group does not have such relevance to improved crop production as do the vesicular-arbuscular and ectotrophic mycorrhizal associations.

The occurrence of mycorrhizal associations in the tropics

The detailed surveys by Janse (1896) and Johnston (1949) recorded only the vesicular-arbuscular type of association. It is noteworthy that the ectotrophic mycorrhizal association has been found only rarely in the natural tropical forest. Redhead (1968a) investigated the incidence of mycorrhizal associations in 51 tree species indigenous to the Lowland Rain Forest of Nigeria and in 15 exotic tree species. All the exotic species and 44 of the indigenous species were found to have endotrophic mycorrhizal associations, and three indigenous species had ectotrophic associations. Thomazini (1974) found a similar proportion of mycorrhizal associations in plants in Brazil. 56 species were found to have endotrophic associations, two ectotrophic, and two ectendotrophic mycorrhizal associations. The almost universal occurrence of the vesicular-arbuscular mycorrhizal association in the tropics has also been confirmed from India (Thapar and Khan 1973) and from the Philippines (Tupas and Sajise 1976).

There are also numerous records of the occurrence of the vesicular-arbuscular mycorrhizal association on important crop plants of the tropics, notably cacao, citrus, coconut, cotton, maize, sweet potato, sugar cane, rubber, tea, tobacco, tomato, and many different species of timber trees, including cypress and teak. In contrast there is only one crop genus recorded with an ectotrophic mycorrhizal association, this is *Pinus*. Wherever pines have been introduced the mycorrhizal association has proved obligatory for growth beyond the nursery stage (Mikola 1970). There is no record of any tropical

crop plant which does not normally develop a mycorrhizal association.

The only records of naturally occurring ectotrophic mycorrhizal
associations are given in Table 16.1.

Table 16.1. Naturally occurring ectotrophic mycorrhizal associations

Species	Country	Author
Caesalpiniaceae		
Afzelia africana	Ghana	Jenik and Mensah (1967)
	Nigeria	Redhead (1968a)
Afzelia bella	Nigeria	Redhead (1960, 1968a)
	Zaire	Fassi and Fontana (1962)
Anthonotha macrophylla	Zaire	Fassi and Fontana (1962)
Bauhinia holophylla	Brazil	Thomazini (1974)
Brachystegia eurycoma	Nigeria	Redhead (1968a)
Brachystegia laurentii	Zaire	Fassi and Fontana (1962)
Gilbertiodendron dewevrei	Zaire	Fassi (1963)
Julbernardia seretii	Zaire	Fassi and Fontana (1961)
Monopetalanthus sp	Zaire	Fassi and Fontana (1962)
Paramacrolobium coeruleum	Zaire	Fassi and Fontana (1962)
Paramacrolobium fragrans	Zaire	Fassi and Fontana (1962)
Dipterocarpaceae		
Anisoptera laevis	Malaya	Singh (1966)
Balanocarpus heimii		
Dipterocarpus oblongifolius		
Dipterocarpus sublamellatus		
Dryobalanops aromatica		
Hopea ferruginea		
Hopea sp		
Shorea curtisii		
Shorea leprosula		
Shorea macroptera		
Shorea ovalis		
Shorea pauciflora		
Vatica papuana		
Euphorbiaceae		
Uapaca togoensis	Nigeria	Redhead (1974)
Fagaceae		
Quercus spicata	Malaya	Singh (1966)
Myrtaceae		
Camponanesia coerulea	Brazil	Thomazini (1974)

The ectotrophic mycorrhizal association also occurs in natural
pine stands. Dr M. H. Ivory of the Commonwealth Forestry Institute,

Oxford, has made collections of mycorrhizal associates of pines from the natural tropical forests of *Pinus caribaea* and *P. oocarpa* in Central America and the Bahamas. These are currently being studied and tested at the Commonwealth Forestry Institute.

The reason for the infrequent occurrence of ectotrophic mycorrhizal associations in the lowland tropics is obscure. It could be because most ectrotrophic mycorrhizal fungi have an optimum temperature for growth far below the ambient temperature of the lowland tropics. In view of the apparent age of the symbiotic state in geological terms, it is hard to imagine that heat tolerant strains could not have evolved. They must occur for instance in *Rhizopogon luteolus* which has spread widely in Australia and which appears to be the most prevalent fungus forming mycorrhizal associations on Nigerian pine introductions.

The identity of tropical mycorrhizal fungi

The fungi forming the vesicular-arbuscular type of mycorrhizal association are species of the Endogonaceae. Janos (1975) identified *Sclerocystis dussii* (Pat.) von Hohn and an *Acaulospora* sp in previously sterilized soil infected with mycorrhizal cacao roots in Costa Rica. Redhead (1974, 1977) examined a range of forest and savannah soils in Nigeria and found numerous species, including *Glomus fasciculatus* (Gerdemann and Trappe 1974) and six different kinds of *Gigaspora*, one of which is possibly *G. gilmorei* (Gerdemann and Trappe 1974), a second is similar to *G. gigantea* (Nicol. and Gerd.) (Gerdemann and Trappe 1974), and a third is an unnamed species described by Old, Nicolson, and Redhead (1973). Spores of a *Sclerocystis* sp were also regularly seen in Nigerian soils. Sanni (1976a,b) has also recorded *Gigaspora gigantea* and other species of Endogonaceae in Nigerian soils. Two mycorrhizal forming species have recently been isolated from forest trees in India by Gerdemann and Bakshi (1976) and named *Glomus multicaulis* Gerdemann and Bakshi and *Sclerocystis sinuosa* Gerdemann and Bakshi.

It is apparent that several species and genera of Endogonaceae have a world-wide distribution. The reason for this wide distribution is not obvious because Endogonaceous spores are so large and appear to have no special mechanism for dispersal other than by small animals which might eat them. The world-wide distribution may be accounted for by the great antiquity of this group of fungi which may have evolved along with the first land plants over 300 million years ago (Nicolson 1975).

Some of the fungi involved in the naturally occurring ectotrophic mycorrhizal associations bear clamp connections and must therefore be Basidiomycetes but the identity of most of the species involved is not known. Several distinctly different species of mycorrhizal fungi must occur on *Afzelia* spp and on *Brachystegia eurycoma* because these fungi have mycelia which are in some cases quite distinct. These fungi can be used in the form of mycorrhizal root fragments to synthesize mycorrhizal associations with *Afzelia africana, A. bipindensis,* and *Brachystegia eurycoma.* It seems that these species of Caesalpiniaceae are disposed to form ectotrophic mycorrhizal associations with a variety of fungi. Potted plants of these species, initially in sterile media, tended to develop mycorrhizal associations from air-borne propagules. Such associations showed some variation and it appeared that more than one species of fungus was involved. Attempts to develop mycorrhizal associations of *Afzelia bella* and *B. eurycoma* using inocula of *Pinus caribaea* roots bearing a mycorrhizal association formed by *Rhizopogon luteolus* were not successful, neither was it possible to produce mycorrhizal associations on *P. caribaea* using mycorrhizal roots of *Afzelia* spp and *B. eurycoma* (Redhead 1974). If this had proved successful it would have had important implications for the introduction of pines into new areas.

Redhead (1974) attempted to obtain pure cultures of the fungi forming associations on *Afzelia* spp and *B. eurycoma* but without success. Fructifications of an *Inocybe* sp grew adjacent to experimental mycorrhizal *Afzelia bella* seedlings on four occasions and the mycelium at the base of the stipe appeared identical to that of the mycorrhizal sheath (Redhead 1968b). Fassi (personal communication) also reported fructifications of an *Inocybe* sp associated with a *Gilbertiodendron* sp in the forests of Zaire. Attempts to obtain a culture from the Nigerian *Inocybe* sporophore were not successful.

Pines introduced to Africa have sometimes been inoculated by the introduction of soil containing root fragments from older pine stands, introduced contrary to quarantine regulations. Perhaps this is why the mycorrhizal association appears to have developed spontaneously on pine in some countries. Amongst fungi identified on pines in the tropics are *Suillus (Boletus) granulatus* (L. ex Fr.) O. Kuntze in Zaire (Thoen 1974), *S. granulatus* and *Coltricia cinnamomea* in West Malaysia (Ivory 1975), *Rhizopogon luteolus* Fr. and Nordh. in Nigeria (Momoh 1976), *Boletus luteus* L., *Scleroderma bovista* (Fr.), and *Hebeloma crustiliniforme* (Fr.) Quel. in Kenya (Gibson 1963), and a *Rhizopogon* sp in Sri Lanka (Mikola, personal communication).

Dr M. H. Ivory (personal communication) has recently identified several fungi from the natural *Pinus caribaea* and *P. oocarpa* forests of Central America and the Bahamas and confirmed that they will form symbiotic associations with pine seedlings in the laboratory. These fungi include the following: *Gyroporus castaneus, Pisolithus tinctorius, Rhizopogon nigrescens, Scleroderma (geaster)?, Suillus cothurnatus, Suillus cf. Hololeucus, Thelephora* sp.

The effect of the mycorrhizal association on plant growth under tropical conditions

Vesicular-arbuscular mycorrhizal associations

There has been little experimental work carried out in the tropics on the effect of the vesicular-arbuscular (VA) mycorrhizal association on plant growth. Janos (1975) found that mycorrhizal inoculation almost doubled the height growth of three tropical rain forest trees but had little effect on a fourth species (Table 16.2).

Table 16.2. Effect of VA mycorrhiza on lowland rain forest trees in Costa Rica (from Janos 1975)

Species	Mean height (cm) Treatment			No. of weeks after inoculation
	1	2	3	
Inga oerstediana	12.3	11.5	19.4	25
Sickingia maxonii	13.2	13.4	14.6	36
Vitex cooperi	6.5	7.4	16.5	24
Unidentified sp of Euphorbiaceae	4.8	5.2	8.3	23

Key to treatments
1. Sterilized soil.
2. Sterilized soil + microbial filtrate + sterilized diced *Cacao* roots.
3. Sterilized soil + diced *Cacao* roots.

Redhead (1975) in Nigeria found that inoculated *Khaya grandifoliola* which developed a heavy incidence of the association produced over six times as much dry matter as *K. grandifoliola* which had only a slight or no incidence of the association. Sanni (1976a), using a similar inoculum in Nigeria, found that it increased the growth of cowpea, maize, and tomato, especially with the addition of $Ca_3(PO_4)_2$. These are sufficient examples to suggest that the vesicular-arbuscular mycorrhizal association is at least as important for the growth of plants in the tropics as it is for plants growing in the subtropics and temperate regions.

Ectotrophic mycorrhizal associations

The association is generally accepted as beneficial and roots in symbiosis with the fungus are more effective nutrient absorbing systems (Harley 1969). The association is particularly important for pines. If they do not form a mycorrhizal association the plants do not grow beyond the nursery stage but become chlorotic and gradually die when planted out in the field. For this reason inocula of ectotrophic mycorrhizal fungi have been introduced into many parts of the tropics with the intention of encouraging the growth of pines (Briscoe 1969; Hacskaylo and Vozzo 1959; Vozzo and Hacskaylo 1971) and Mikola (1970) has reviewed the history of the introduction of species of pine and their inocula into various territories of the tropics.

Studies on pines in temperate regions indicate that low levels of available nitrogen and phosphorus are necessary for the ectotrophic mycorrhizal association to develop but increases much above these levels depress the development of the association (Hatch 1937; Björkman 1942). In Britain, coniferous stock may not develop the association in nurseries which have been heavily manured with hop waste or farmyard manure, but associations develop when the young plants are planted out on forest soils, as the symbiotic fungi are ubiquitous in temperate forest areas.

Information on the importance of the association for broad-leaved species in the tropics is limited. Redhead (1974) carried out research on the mycorrhizal associations of *Brachystegia eurycoma* Harms. This tree is the commonest species of Caesalpiniaceae in the lowland rain forest of Nigeria and forms ectotrophic mycorrhizal associations with more than one species of fungus. Experiments were carried out on *Brachystegia* using two different mycorrhizal fungi to study whether the association increased the growth of the tree and whether the addition of nitrogen and phosphorus affected the development of the association.

The roots of *Brachystegia* bearing two distinctly different mycorrhizal fungi from separate sites were used as inocula. One was a white fungus, the other brown.

The white mycorrhizal association.   All the fine rootlets were covered by a fungal mantle 40 μm in average thickness and composed of fine, almost colourless hyphae. No root hairs were observed; some mycorrhizal rootlets appeared smooth, others bore extensive masses of loose hyphae and numerous mycelial strands up to 70 μm thick. *En masse* such rootlets had a distinctive smell. This association was similar to genus Ca of the Dominik classification (Dominik 1959).

At a distance of 0.5 cm from the root tip the fine mycorrhizal rootlets varied from 200–750 μm in diameter, of which a third was composed of the mantle of fine, compact, hyaline hyphae 2–3 μm in diameter. These penetrated between the cortical cells, which formed a layer often only one cell deep around the endodermis, but were not observed to penetrate the root cells. The hyphae were septate but clamp connections were not observed.

The brown mycorrhizal association.   The fine rootlets showed a very conspicuous mycorrhizal association. In nature all rootlets were affected but where the fungus was introduced to a pot-grown seedling raised under sterile conditions, although the fungal spread was very rapid, it was possible to obtain uninfected rootlets growing side by side with mycorrhizal rootlets. Mycorrhizal rootlets stopped elongation and were very short in comparison with uninfected rootlets. The mantle was 40 μm in average thickness and outgrowing hyphae radiated densely, occasionally forming loose mycelial strands, and often binding the soil and adjacent rootlets together. The hyphae were medium brown, 3–4 μm in diameter, and showed prominent clamp connections. The association was similar to genus Gb of the Dominik classification.

The fine mycorrhizal rootlets were on average 300–500 μm in diameter, at least a third of which was a mantle made up of two zones, an outer zone of large 'cells' up to 10 μm in diameter and an inner zone of pseudoparenchyma approximately 4 μm in diameter. The root cortex comprised one or two layers of large cells surrounding a narrow layer of disorganized brownish cells outside the endodermis. Hyphae penetrated between the cortical cells as far as the brownish layer and occasionally entered the cells of the cortex, showing constrictions where they passed through the cell wall.

Seedlings were inoculated with these mycorrhizal fungi and grown at two levels of nutrients, at very low levels of nitrogen and phosphorus and at very high levels of nitrogen. The mycorrhizal plants showed significantly better growth than the uninoculated control plants (Tables 16.3 and 16.4).

Both white and brown mycorrhizal associations significantly increased the height and dry weights of the leaves, stem, and roots of *B. eurycoma*. The weight of the mycorrhizal mycelium and rootlet mantle could be expected to increase the root weight and root:shoot ratio compared with that of non-mycorrhizal plants but, as the weights of the leaves and stem were also greatly increased, the mycorrhizal association obviously confers considerable advantage to this species. This could be expected to give a competitive advantage to *B. eurycoma* and might explain why this species is so common in low-

Table 16.3. Mean stem heights and dry weights of *Brachystegia eurycoma* grown at low levels of nutrients after inoculation with two mycorrhizal fungi

Treatment		Stem height (cm)	Mean dry weight (g)			
			Leaves	Stem	Root	Total
Inoculation sources	No inoculation	30.5	0.76	1.94	1.60	4.30
All fertilizer levels combined	*B. eurycoma* White mycorrhiza	30.5	0.87	2.10*	2.14†	5.11†
	B. eurycoma Brown mycorrhiza	32.9	0.86	2.58*	2.33†	5.77†
Fertilizer treatment	No added nutrient	27.3	0.66	1.61	1.58	3.85
	6 p.p.m. N + 1 p.p.m. P	31.9	0.79	2.06	1.75	4.60
All inocula combined	6 p.p.m. N + 5 p.p.m. P	29.8	0.77	1.94	2.09	4.80
	6 p.p.m. N + 25 p.p.m. P	36.0†	1.07†	3.05†	2.55†	6.67†

*Difference significant at the 5 per cent level.
†Difference significant at the 1 per cent level.

land rain forest and riverain forest in Nigeria. It would be interesting to ascertain whether the other species of *Brachystegia* and the closely related *Julbernardia* spp, which dominate the 'Miombo' woodland south of the equatorial forest block of central Africa, also form ectotrophic mycorrhizal associations.

There was no lack of mycorrhizal development on the *B. eurycoma* which received minimal amounts of nutrients, nor did development seem in any way reduced by applications of nitrogen at levels high enough to prove harmful to the seedlings. This is in contrast to the findings of many workers on *Pinus* spp who found that certain low levels of nutrient were needed for a mycorrhizal association to develop but that high concentrations of available nitrogen reduced or inhibited the formation of mycorrhizal associations (Hatch 1937; Björkman 1942; Fowells and Krauss 1949; Hacskaylo and Snow 1959). Further studies on the physiology of nutrition of tropical ectotrophic mycorrhizal trees are advisable as the results may have important implications for forestry practice.

Absorption of water and soil nutrients

The extensive hyphal growth of both vesicular-arbuscular and ectotrophic mycorrhizal fungi increases the absorbing area of the roots manyfold. Wetsieving of even the driest savannah soils will usually

Table 16.4. Mean stem heights, dry weights, and nitrogen content of *Brachystegia eurycoma* grown at three levels of nitrogen and two levels of potassium after inoculation with two mycorrhizal fungi

Treatment	Stem height (cm)	Mean dry weight (g)				Nitrogen content Dry matter (%)		
		Leaves	Stem	Roots	Total	Leaves	Stem	Roots
Inoculation sources								
All fertilizer levels combined								
No inoculation	22.4	0.74	1.07	1.10	2.91	2.2	0.8	0.9
White mycorrhiza	26.3*	1.23†	1.65†	2.05†	4.93†	2.1	0.8	0.9
Brown mycorrhiza	28.6*	1.54†	1.94†	2.08†	5.56†	2.2	0.7	0.8
Fertilizer treatments	†	†	†	†	†	†	†	†
Total daily amount per plant.								
All fungal inocula combined								
No added nutrient	18.3	0.58	1.00	1.21	2.79	1.7	0.5	0.6
0.0016 g N	23.3	0.98	1.24	1.70	3.92	1.9	0.7	0.7
0.0024 g N + 0.0011 g K	28.2	1.41	2.41	2.74	6.56	2.1	0.7	0.7
0.0040 g N + 0.0011 g K	30.5	1.46	2.16	2.75	6.37	1.9	0.5	0.8
0.0112 g N	23.0	1.01	0.91	0.79	2.71	2.6	1.1	1.2
0.0120 g N + 0.0011 g K	31.1	1.57	1.62	1.29	4.48	2.7	1.0	1.3

*Difference significant at the 5 per cent level.
†Difference significant at the 1 per cent level.

yield large amounts of typical '*Endogone*' mycelium. This appears quite fresh, even when annual herbs have completely withered. Inspection of mycorrhizal roots of *Afzelia* spp, *Brachystegia eurycoma*, and *Pinus caribaea* show extensive mycelia ramifying far beyond the limits of the absorbing roots, although *P. caribaea* mycelium appears to dry out and wither under plantation conditions during the hot, tropical dry season.

In all but true tropical rain forest, water is a limiting factor for growth during several months of the year. The increased efficiency of an '*extended*' root system is likely to assist greatly those plants with a well developed mycorrhizal association. Moreover, there is considerable evidence that mycorrhizal hyphae are very efficient absorbers of ions from the soil solution. This appears particularly important in the case of phosphorus which is deficient in many tropical soils. The vesicular-arbuscular mycorrhizal hyphae do not seem able to actively dissolve phosphates but they can withdraw the ions from the soil solution even at very low concentrations. This results in a continuous net movement into solution even from such relatively insoluble phosphorus minerals as calcium phosphate, $Ca_3(PO_4)_2$, and strengite, $Fe(OH_2)H_2PO_4$ (Sanni 1976b). The role of vesicular-arbuscular mycorrhizal associations in plant nutrition has been reviewed by Mosse (1973).

Resistance to disease

Marx (1969a,b) and Marx and Davey (1969a,b) demonstrated the antagonism of some ectotrophic mycorrhizal fungi to certain root pathogenic fungi. Studies on the vesicular-arbuscular mycorrhizal association indicated that the association protects tobacco roots against *Thielaviopsis basicola* (Baltruschat and Schönbeck 1972) but Ross (1972) found that the association increased *Phytophthora* root-rot of soya beans. Further research is needed on these important aspects of the ecology of the association.

Trends in research on the mycorrhiza of natural tropical plants

So far research in the tropics has been largely exploratory. There is scope to extend this exploratory work to identify the principally occurring mycorrhizal fungi and to relate them to the ecology of the plants with which they associate. For example, the 'Miombo' woodlands are one of the world's most extensive vegetation zones; stretching from Tanzania they extend a thousand miles southwards through Zambia and Mozambique into Zimbabwe and westwards to Angola on the Atlantic Ocean. Is it a coincidence that this vegetation is dominated by species of Caesalpiniaceae, notably *Brachystegia* spp and the closely related *Julbernardia* spp? Perhaps these species also form ectotrophic mycorrhizal associations. It would be of considerable ecological interest to find out.

Research into the mycorrhizal associations of natural tropical forests has important implications for agriculture and forestry. The ecological preference of naturally occurring mycorrhizal fungal species and strains, with respect to conditions of temperature, moisture, soil pH, and nutrient status, must be determined as a preliminary to building up stocks of valuable strains. Research in the subtropics has indicated that certain heat-tolerant strains of *Gigaspora* spp may be especially well suited to the cultivation of summer crops in Florida, USA (Schenck and Schroder 1974; Schenck, Graham, and Green 1975). The screening of identified fungal strains with major crop plants may indicate that some combinations have particular advantage, either for normal growth or under special conditions. Mosse (1972a,b) has already demonstrated this for the tropical grass *Paspalum notatum* and has reviewed past experience of the specificity of mycorrhizae (Mosse 1975).

Studies on mycorrhizae have developed rapidly in recent years and two fields appear particularly important for the tropics:

Investigation into methods of inoculum storage and propagation

The feasibility of international co-operation in the supply of inocula of mycorrhizal fungi depends on effective methods of propagule storage and transport, and on better techniques of inoculation and multiplication. The spores of vesicular-arbuscular mycorrhizal fungi can be multiplied and used as an effective inoculum even though the fungi have not been grown in pure culture (Mosse 1959; Hattingh and Gerdemann 1975). Ectotrophic mycorrhizal fungi have usually proved difficult to introduce from pure cultures. One of the best documented introductions of such mycorrhizal fungi was in Puerto Rico (Briscoe 1959; Vozzo 1966; Hacskaylo and Vozzo 1967; Vozzo and Hacskaylo 1971). Gibson (1963) used dried sporophores successfully in East Africa and more recently spores have been successfully stored after freeze-drying and used after storage for three months (Theodorou and Bowen 1973). Preliminary studies have also been carried out on the freeze-drying of vesicular-arbuscular mycorrhizal inoculum (Crush and Pattison 1975). Lamb and Richards (1974) have also demonstrated that basidiospores and oidia showed high heat tolerance and germinated well after sixty days. The use of seed pelleted with the appropriate mycorrhizal fungus is an attractive proposition which deserves further study in the tropics.

Investigation of the effects of herbicides, insecticides, fungicides, and soil sterilants on mycorrhizal plants in nurseries and under various field conditions

The use of chemicals in plant protection and weed control is becoming more prevalent in the tropics and is likely to increase as labour becomes more expensive. The possible consequences should be recognized and this requires research under field conditions in order to avoid such catastrophes as was experienced with *Citrus* in Florida where germinated seedlings failed to grow in the absence of mycorrhiza (Kleinschmidt and Gerdemann 1972; Schenck and Tucker 1974; Jalali and Domsch 1975).

These studies could be carried out in controlled environments in temperate regions but wherever possible this work should be done in the tropics. The development of these important programmes would ensure that facilities are strengthened in the tropical territories and that indigenous staff are trained under tropical conditions. This is essential for the long-term benefit of agriculture and forestry in the tropics.

References

Baltruschat, H. and Schönbeck, F. (1972). Untersuchungen über den Einfluss der endotrophen Mycorrhiza auf die Chlamydosporenbildung von *Thielaviopsis basicola* in Tabakwurzeln. *Phytopath. Z.* 74, 358-61.

Björkman, E. (1942). Über die Bedingungen der Mykorrhizabildung bei Kiefer und Fichte. *Symb. bot. upsal.* 6, 1-190.

Briscoe, C. B. (1959). Early results of mycorrhizal inoculation of pine in Puerto Rico. *Carib. For.* 20, 73-7.

Crush, J. R. and Pattison, A. C. (1975). Preliminary results on the production of vesicular-arbuscular mycorrhizal inoculum by freeze-drying. In *Endomycorrhizas* (eds. F. E. Sanders, B. Mosse, and P. B. Tinker) pp. 485-93. Academic Press, London.

Dominik, T. (1959). Synopsis of a new classification of the ectotrophic mycorrhizae established on morphological and anatomical characteristics. *Mycopathologia* 11, 359-67.

Fassi, B. (1963). Die Verteilung der ektotrophen Mykorrhizen in der Streu und in der oberen Bodenschicht der *Gilbertiodendron dewevrei* (Caesalpiniaceae) —Wälder in Kongo. In *Mykorrhiza.* Proc. Intern. Mykorrhiza-symposium, Weimar, 1960 (eds. W. Rawald and H. Lyr) pp. 297-302. G. Fischer, Jena.

—— and Fontana, A. (1961). Le micorrize ectotrofiche di *Julbernardia seretii*, Caesalpinaceae del Congo. *Allionia* 7, 131-51.

—— —— (1962). Micorrize ectotrofiche di *Brachystegia laurentii* e di alcune altre Caesalpiniaceae minori del Congo. *Allionia* 8, 121-31.

Fowells, H. A. and Krauss, R. W. (1949). The inorganic nutrition of Loblolly pine and Virginia pine with special reference to nitrogen and phosphorus. *Forest Sci.* 5, 95-112.

Frank, A. B. (1885). Über die auf Wurzelsymbiose beruhende Ernährung gewisser Bäume durch unterirdische Pilze. *Ber. dt. bot. Ges.* 3, 128-45.

Gerdemann, J. W. (1968). Vesicular-arbuscular mycorrhiza and plant growth. *A. Rev. Phytopathol.* 6, 397-418.

— and Bakshi, B. K. (1976). Endogonaceae of India: two new species. *Trans. Br. mycol. Soc.* 66, 340-3.

— and Trappe, J. M. (1974). The Endogonaceae in the Pacific Northwest. *Mycologia Memoir*, Vol. 5.

Gibson, I. A. S. (1963). Eine Mitteilung über die Kiefernmykorrhiza in den Wäldern Kenias. In *Mykorrhiza.* Proc. Intern. Mykorrhizasymposium, Weimar, 1960 (eds. W. Rawald and H. Lyr) pp. 49-51. G. Fischer, Jena.

Hacskaylo, E. and Snow, A. G. (1959). Relations of soil nutrients and light to prevalence of mycorrhizae on pine seedlings. Station Paper No. 125, N.E. Forest Exp. Station, Upper Darby, Pa., USA.

— and Vozzo, J. A. (1967). Inoculation of *Pinus caribaea* with pure cultures of mycorrhizal fungi in Puerto Rico. *Proc. XIV IUFRO Congress, Munich, 1967*, Sect. 24, Vol. 5, pp. 139-48.

Harley, J. L. (1968). Fungal symbiosis: Presidential address to the British Mycological Society. *Trans. Br. mycol. Soc.* 51, 1-11.

— (1969). *The biology of mycorrhiza*, 2nd edn. Leonard Hill, London.

Hartig, R. (1888). Die pflanzlichen Wurzelparasiten. *Allg. Forst-u. Jagdztg* 64, 118-23.

Hatch, A. B. (1937). The physical basis of mycotrophy in *Pinus*. *Black Rock Forest Bull.*, Vol. 6.

Hattingh, M. J. and Gerdemann, J. W. (1975). Inoculation of Brazilian sour orange seed with an endomycorrhizal fungus. *Phytopathology* 65, 1013-16.

Ivory, M. H. (1975). Mycorrhizal studies on exotic conifers in West Malaysia. *Malay. Forester* 38, 149-52.

Jalali, B. L. and Domsch, K. H. (1975). Effect of systemic fungitoxicants on the development of endotrophic mycorrhiza. In *Endomycorrhizas* (eds. F. E. Sanders, B. Mosse, and P. B. Tinker) pp. 619-26. Academic Press, London.

Janos, D. P. (1975). Effects of vesicular-arbuscular mycorrhizas on lowland tropical rain forest trees. In *Endomycorrhizas* (eds. F. E. Sanders, B. Mosse, and P. B. Tinker) pp. 437-46. Academic Press, London.

Janse, J. M. (1896). Les endophytes radicaux de quelques plantes Javanaises. *Annls Jard. bot. Buitenz.* 14, 53-212.

Jenik, J. and Mensah, K. O. A. (1967). Root system of tropical trees. I. Ectotrophic mycorrhizae of *Afzelia africana* Sm. *Preslia* 39, 59-65.

Johnston, A. (1949). Vesicular-arbuscular mycorrhiza in Sea Island cotton and other tropical plants. *Trop. Agric., Trin.* 26, 118-21.

Kleinschmidt, G. D. and Gerdemann, J.W. (1972). Stunting of *Citrus* seedlings in fumigated nursery soils related to the absence of endomycorrhizae. *Phytopathology* 62, 1447-53.

Lamb, R. J. and Richards, B. N. (1974). Survival potential of sexual and asexual spores of ectomycorrhizal fungi. *Trans. Br. mycol. Soc.* 62, 181-91.

Marx, D. H. (1969a). Antagonism of mycorrhizal fungi to root pathogenic fungi and soil bacteria. *Phytopathology* 59, 153-63.

— (1969b). Production, identification, and biological activity of antibiotics produced by *Leucopaxillus cerealis* var. *piceina*. *Phytopathology* 59, 411-17.

— and Davey, C. B. (1969a). Resistance of aseptically formed mycorrhizae to infection by *Phytophthora cinnamomi*. *Phytopathology* 59, 549-58.

— — (1969b). Resistance of naturally occurring mycorrhizae to infections by *Phytophthora cinnamomi*. *Phytopathology* 59, 559-65.

Mikola, P. (1970). Mycorrhizal inoculation in afforestation. *Int. Rev. For. Res.* 3, 123-96.

Momoh, Z. O. (1976). Synthesis of mycorrhiza of *Pinus oocarpa. Ann. appl. Biol.* 82, 221-6.

Mosse, B. (1959). The regular germination of resting spores and some observations on the growth requirements of an *Endogone* sp. causing vesicular-arbuscular mycorrhiza. *Trans. Br. mycol. Soc.* 42, 273-86.

— (1963). Vesicular-arbuscular mycorrhiza: an extreme form of fungal adaptation. In *Symbiotic associations* (eds. P. S. Nutman and B. Mosse) pp. 146-70. Cambridge University Press.

— (1972a). The influence of soil type and *Endogone* strain on the growth of mycorrhizal plants in phosphate deficient soils. *Rev. Ecol. Biol. Sol.* 9, 529-37.

— (1972b). Effects of different *Endogone* strains on the growth of *Paspalum notatum. Nature, Lond.* 239, 221-3.

— (1973). Advances in the study of vesicular-arbuscular mycorrhiza. *A. Rev. Phytopathol.* 11, 171-96.

— (1975). Specificity in VA mycorrhizas. In *Endomycorrhizas* (eds. F. E. Sanders, B. Mosse, and P. B. Tinker) pp. 469-84. Academic Press, London.

Nicolson, T. H. (1975). Evolution of vesicular-arbuscular mycorrhizas. In *Endomycorrhizas* (eds. F. E. Sanders, B. Mosse, and P. B. Tinker) pp. 25-34. Academic Press, London.

Old, K. M., Nicolson, T. H., and Redhead, J. F. (1973). A new species of mycorrhizal *Endogone* from Nigeria with a distinctive spore wall. *New Phytol.* 72, 817-23.

Redhead, J. F. (1960). A study of mycorrhizal associations in some trees of Western Nigeria. Diploma Thesis, Oxford University.

— (1968a). Mycorrhizal associations in some Nigerian forest trees. *Trans. Br. mycol. Soc.* 51, 377-87.

— (1968b). *Inocybe* sp. associated with ectotrophic mycorrhiza on *Afzelia bella* in Nigeria. *Commonw. For. Rev.* 47, 63-5.

— (1974). Aspects of the biology of mycorrhizal associations occurring on tree species in Nigeria. Ph.D. Thesis, University of Ibadan, Nigeria.

— (1975). Endotrophic mycorrhizas in Nigeria: some aspects of the ecology of the endotrophic mycorrhizal association of *Khaya grandifoliola* C.DC. In *Endomycorrhizas* (eds. F. E. Sanders, B. Mosse, and P. B. Tinker) pp. 447-59. Academic Press, London.

— (1977). Endotrophic mycorrhizas in Nigeria: species of the Endogonaceae and their distribution. *Trans. Br. mycol. Soc.* 69, 275-80.

Richards, P. W. (1952). *The tropical rain forest.* Cambridge University Press.

Ross, J. P. (1972). Influence of Endogone mycorrhiza on *Phytophthora* rot of soybean. *Phytopathology* 62, 896-7.

Sanni, S. O. (1976a). Vesicular-arbuscular mycorrhiza in some Nigerian soils and their effect on the growth of cowpea (*Vigna unguiculata*), tomato (*Lycopersicon esculentum*) and maize (*Zea mays*). *New Phytol.* 77, 662-71.

— (1976b). Vesicular-arbuscular mycorrhiza in some Nigerian soils: the effect of *Gigaspora gigantea* on the growth of rice. *New Phytol.* 77, 673-4.

Schenck, N. C. and Schroder, V. N. (1974). Temperature response of *Endogone* mycorrhiza on soybean roots. *Mycologia* 64, 600-5.

— and Tucker, D. P. H. (1974). Endomycorrhizal fungi and the development of citrus seedlings in Florida fumigated soils. *J. Am. Soc. Hort. Sci.* 99, 284-7.

— Graham, S. O., and Green, N. E. (1975). Temperature and light effect on contamination and spore germination of vesicular-arbuscular mycorrhizal fungi. *Mycologia* 67, 1189-92.

Singh, K. G. (1966). Ectotrophic mycorrhiza in equatorial rain forest. *Malay. Forester* 39, 13-19.

Thapar, H. S. and Khan, S. N. (1973). Studies on endomycorrhiza in some forest species. *Proc. Indian Nat. Sci. Acad. B.* Forest Research Institute, Dehra Dun, India.

Theodorou, C. and Bowen, G. D. (1973). Inoculation of seeds and soil with basidiospores of mycorrhizal fungi. *Soil Biol. Biochem.* 5, 765–71.

Thomazini, L. I. (1974). Mycorrhiza in plants of the 'Cerrado'. *Pl. Soil* 41, 707–11.

Thoen, D. (1974). Preliminary data on mycorrhizas and mycorrhizal fungi in plantations of exotics of Upper Shaba (Republic of Zaire). *Bull. Rech. agron. Gembloux* 9, 215–27.

Treub, M. (1885). Onderzoekingen over Sereh-zick Suikerriet. *Meded. Pl. Tuin. Batavia II.*

Tupas, G. L. and Sajise, P. E. (1976). Mycorrhizal associations in some savanna and reforestation trees. *Kalikasan* 5, 235–40.

Vozzo, J. A. (1966). Inoculation of pine with mycorrhizal fungi in Puerto Rico. Ph.D. Thesis, George Washington University, USA.

—— and Hacskaylo, E. (1971). Inoculation of *Pinus caribaea* with ectomycorrhizal fungi in Puerto Rico. *Forest Sci.* 17, 239–45.

Wakker, J. H. and Went, F. A. F. C. (1898). *Die Ziekten van het Suikerriet op Java.* Leiden.

17 Preliminary notes on mycorrhizae in a natural tropical rain forest

S. Riess and A. Rambelli

Research on mycorrhizal forms present in a natural rain forest has been carried out within the Tai Project entitled 'Effect of increasing human activities on a south western Ivory Coast tropical rain forest'.

Orchid mycorrhizae

Mycorrhizae of epiphytic orchids were studied in three stages. In the first morphological examinations were made on the symbiotic structures which correspond to those typically observed in orchid mycorrhizae (Bernard 1904, 1909; Burgeff 1909, 1943, 1959; Fuchs and Ziegenspeck 1925). The second stage concerned the isolation of the fungal symbiont of *Aerangis biloba*. The techniques were derived from Freccero and Fanelli (1975), Burgeff (1932, 1959), and Harvais and Hadley (1967). Surface sterilization was carried out with 30 per cent H_2O_2 for 5–8 minutes. In the third stage reinoculation trials were made with the mycelia obtained in the second stage. Mycelia to be tested were grown on a natural medium,* to which were transplanted protocorms of *Cattleya* obtained from meristem culture. Positive results were obtained.

Morphological studies on tree mycorrhizae

In the first stage of our study no attempt was made to identify the root tips collected, but a screening was made of the mycorrhizal forms which could be found in the Tai forest. Two research programmes are now carried out: determination of mycorrhizal forms of identified root tips, and collection and identification of ndogonaceous spores in the soil.

In the first stage a total of 270 root tips was collected at three different periods, from at least ten sampling points for each period and over an area of 500 m². Root tips were collected in the upper 5 cm of soil, washed, and placed in a solution of glutaraldehyde at pH 7.2,†

---

*Natural medium A: coniferous bark (250 g), vermiculite (180 g), starch (21 g), and distilled water (700 ml). pH before autoclaving: 6.8–7.0.

†Glutaraldehyde (25 per cent solution) (7.2 ml), phosphate buffer (pH 7.2) (15 ml), distilled water (37.8 ml). Total 60 ml.

which was found to be the best fixing agent, altering neither colour nor structure.

Mycorrhizal forms observed were grouped into four main types; two of them are ectomycorrhizae and the other two VA mycorrhizae. VA mycorrhizae are about ten times more frequent than ectomycorrhizae. The ectomycorrhizae could be divided into two groups: ectendomycorrhizae and true ectomycorrhizae.

The first type of VA mycorrhiza was observed in the majority of root tips examined. The following description is based on observations made on longitudinal sections. Generally speaking, fungal infection spreads into the whole root tip. Fungal hyphae are mainly intracellular and may be seen from the epidermis up to the inner layers of parenchyma. Penetration points and external hyphae may also be seen. The mycelium is of the phycomycetous type, aseptate, with few pseudosepta. Hyphae are thick-walled, from 4.2 to 8.4 μm in diameter, mostly hyaline, sometimes light brown; the cytoplasmic content is mostly homogeneous, rarely granular. Hyphal behaviour is mostly sigmoid, mainly in the inner layers. 'Pelotons', from loose to compact, are clearly evident in the inner layers. Vesicles are the most interesting structures observed; they are frequent, oval or rarely globose in shape, terminal, and intracellular. Intercellular vesicles are rare. Vesicles are always hyaline, double-walled, and their cytoplasmic content is homogeneous; sometimes, however, structures resembling toruloid septate hyphae may be observed in the vesicles. Oval vesicles measure 58.8–47.6 $\times$ 33.6–22.4 μm. Globose vesicles measure 27.3–22.4 μm in diameter.

Sometimes two different mycorrhiza-forming fungi may be observed in the same root tip. They are clearly distinguishable by their different colours and hyphal diameters. The two fungi occupy separate areas of the root tip. Such a situation may possibly present itself at the beginning of mycorrhiza formation.

A typical example of this form of VA mycorrhiza is found in *Drypetes gilgiana* and *Diospyros mannii*. The latter, however, forms fewer mycorrhizal associations.

The second type of endomycorrhiza is less frequent, at least in the periods of the year under consideration. The following description is based on observations made on longitudinal sections of roots. Fungal hyphae penetrate through the epidermis. External hyphae may be seen although infrequently. The mycelium is of the phytomycetous type, rarely pseudoseptate and markedly sigmoid in behaviour, from hyaline to light brown in colour. Hyphae are thick-walled, 2.8–7.8 μm in diameter. Cellular content is always granular. Fungal infection begins with formation of hyphal 'pelotons' which may be observed in the epidermal cells and in the first layers of the parenchymatous tissue. In the inner layers of cortex the main hyphae are intercellular, and intracellular arbuscules and 'sporangioles' are soon digested leaving granular brown residues in the cells. Vesicles are typically formed, but to a lesser extent than in the first type of mycorrhizae; they are roundish, terminal, intracellular, and 19.6–25.4 μm in diameter. Their cytoplasmic content is granular.

References

Bernard, N. (1904). Recherches experimentales sur les Orchidées. *Revue gén. Bot.* XVI, 405–51.

— (1909). L'evolution dans la symbiose. *Annls Sci. nat. (Bot.)* 9 (Ser. 9), 1–196.

Burgeff, H. (1909). *Die Wurzenpilze der Orchideen.* Jena.

— (1932). *Saprophytismus und Symbiose.* Jena.

— (1943). Problematik der Mycorrhiza. *Naturwissenschaften* 47/48, 448–67.

— (1959). Mycorrhiza of orchids. In *The orchids scientific studies*, pp. 361–95.

Freccero, V. and Fanelli, C. (1975). Isolamento dei funghi micorrizogeni del *Pinus radiata.* II. *Pubbl. Cent. Sper. agric. for.* XIII, 29–34.

Fuchs, A. and Ziegenspeck, H. (1925). Bau und Form der Wurzeln der einheimischen Orchideen in Hinblick auf ihre Aufgaben. *Bot. Arch.* 12, 290–379.

Harvais, G. and Hadley, G. (1967). The relation between host and endophyte in Orchid mycorrhiza. *New Phytol.* 66, 205–15.

18 A survey of mycorrhizae in some forest trees of Sri Lanka

D. P. de Alwis and K. Abeynayake

Little work has been carried out on the mycorrhizae of the indigenous flora of Sri Lanka. The aim of the present investigation is to rectify this by surveying the mycorrhizae of some of the indigenous forest trees. The work was carried out at the Kottawa Arboretum, a lowland tropical rain forest located in the wet zone of Sri Lanka at an elevation of 30–60 m having an average annual rainfall of about 3720 mm and a red-yellow podzolic soil (pH between 5.0 and 6.0).

Root material of 63 tree species belonging to twenty-six families were examined. The type of mycorrhizae present in the tree species is given in Table 18.1. Most of the species examined were mycorrhizal (59 out of 63)—the majority having endomycorrhizae of the vesicular-arbuscular type. Five species, viz. *Shorea affinis, Dipterocarpus zeylanicus, Dipterocarpus hispidus, Cotylelobium scarbriusculum,* and *Hopea jucunda,* which showed the presence of ectomycorrhizae were restricted to the family Dipterocarpaceae. The results of the distribution of mycorrhizae are in agreement with the work of Mosse (1963), Meyer (1973), and Singh (1966).

Table 18.1. Types of mycorrhizae present in the lowland tropical rain forest tree species at Kottawa

Family	Species	Type of mycorrhiza
Anacardiaceae	*Campnosperma zeylanica*	endomycorrhiza
	Mangifera zeylanica	endomycorrhiza
	Semecarpus gardneri	endomycorrhiza
	S. subpeltata	endomycorrhiza
Annonaceae	*Cycthocalyx zeylanicus*	endomycorrhiza
	Xylopia championii	endomycorrhiza
Apocynaceae	*Alstonia scholaris*	endomycorrhiza
Burseraceae	*Canarium zeylanicum*	endomycorrhiza
Celastraceae	*Kurrima ceylanica*	endomycorrhiza
Cornaceae	*Mastixia tetrandra*	endomycorrhiza
Dilleniaceae	*Dillenia retusa*	endomycorrhiza
	Schumacheria castaneifolia	endomycorrhiza
	Wormia triquetra	endomycorrhiza
Dipterocarpaceae	*Cotylelobium scarbriusculum*	ectomycorrhiza
	Dipterocarpus hispidus	ectomycorrhiza
	D. zeylanicus	ectomycorrhiza
	Hopea jucunda	ectomycorrhiza
	Shorea affinis	ectomycorrhiza

Family	Species	Type of mycorrhiza
Elaeocarpaceae	*Elaeocarpus subvillosus*	endomycorrhiza
Euphorbiaceae	*Agrostistachys hookeri*	endomycorrhiza
	Antidesma pyrifolium	endomycorrhiza
	Aporosa cardiosperma	endomycorrhiza
	Bridelia moonii	endomycorrhiza
	Chaetocarpus coriaceus	endomycorrhiza
	Ostodes zeylanica	endomycorrhiza
Flacourtiaceae	*Hydnocarpus octandra*	endomycorrhiza
Guttiferae	*Calophyllum bracteatum*	endomycorrhiza
	C. inophyllum	endomycorrhiza
	C. pulcherrimum	endomycorrhiza
	Garcinia cambogia	endomycorrhiza
	G. spicata	endomycorrhiza
	G. terpnophylla	endomycorrhiza
	Mesua thwaitesii	endomycorrhiza
Icacinaceae	*Urandra apicalis*	endomycorrhiza
Lauraceae	*Cryptocarya membranacea*	endomycorrhiza
	Neolitsea cassia	endomycorrhiza
Melastomataceae	*Memecylon arnottianum*	endomycorrhiza
	M. rhinophyllum	endomycorrhiza
Meliaceae	*Amoora rohituka*	endomycorrhiza
Moraceae	*Artocarpus nobilis*	endomycorrhiza
Myristicaceae	*Horsfieldia irya*	endomycorrhiza
	H. iryaghedhi	endomycorrhiza
	Myristica dactyloides	endomycorrhiza
Myrtaceae	*Syzygium aqueum*	endomycorrhiza
	S. makul	endomycorrhiza
Palmae	*Areca catechu*	—
	Caryota urens	endomycorrhiza
Rhizophoraceae	*Anisophyllea cinnamomoides*	endomycorrhiza
	Carallia brachiata	endomycorrhiza
	C. calycina	endomycorrhiza
Rubiaceae	*Byrsophyllum ellipticum*	endomycorrhiza
	Canthium dicoccum	endomycorrhiza
	Randia gardneri	endomycorrhiza
	Timonius jambosella	endomycorrhiza
	Tricalysia erythrospora	endomycorrhiza
Sapotaceae	*Madhuca fulva*	—
	Palaquium grande	endomycorrhiza
	P. pauciflorum	endomycorrhiza
	P. rubiginosum	—
	P. thwaitesii	—
Symplocaceae	*Symplocos coronata*	endomycorrhiza
Thymelaeaceae	*Gyrinopus walla*	endomycorrhiza
Verbenaceae	*Vitex pinnata*	endomycorrhiza

In *Shorea affinis* two types of ectomycorrhizae were seen. The first type (Fig. 18.1a), had a pyramidally pinnate structure and a pink or brown colour, while the second was a nodular or tuberculate type, brown in colour (Fig. 18.1b). *Dipterocarpus zeylanicus* had white or yellow or brown or black monopodial mycorrhizae (Fig. 18.2a) and pinkish brown coralloid mycorrhizae (Fig. 18.2b). *Dipterocarpus hispidus* had yellow, brown or black pyramidally pinnate mycorrhizae (Fig. 18.2c). *Cotylelobium scarbriusculum* had brown monopodial mycorrhizae (Fig. 18.3a). *Hopea jucunda* had monopodial mycorrhizae of a black or brown colour (Fig. 18.3b) and coralloid mycorrhizae of a pinkish brown colour (Fig. 18.3c).

Attempts were made to isolate fungal symbionts from the ectomycorrhizae using 0.1 per cent $HgCl_2$ and 30 vol. H_2O_2 as sterilants and three media—Melin's nutrient solution as modified by Norkans

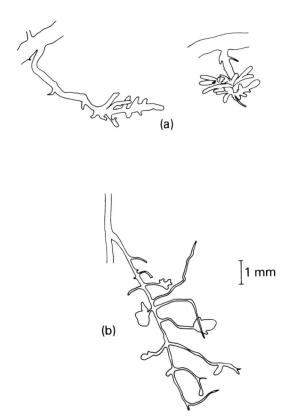

Fig. 18.1. Ectomycorrhizae of *Shorea affinis*. (a) Pyramidally pinnate type. (b) Nodular or tuberculate type.

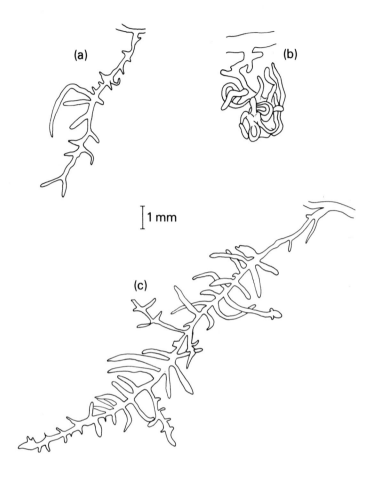

Fig. 18.2. Ectomycorrhizae of *Dipterocarpus*. (a) Monopodial type of *Dipterocarpus zeylanicus*. (b) Coralloid type of *Dipterocarpus zeylanicus*.

with 2 per cent Noble agar, Hagem medium, and modified Fries' medium (Lewis 1961). For surface sterilization of ectomycorrhizae the most suitable sterilant was 0.1 per cent $HgCl_2$. The best isolation of fungi from ectomycorrhizae was in Fries' modified medium. Even though mycorrhizae of different forms and colour were used for the isolation, only the brown and black mycorrhiza yielded isolates. No successful isolation was possible with the white, pinkish, or yellow forms. The brown tuberculate or nodular mycorrhiza of *Shorea affinis* did not yield any isolates. Successful isolations were made from 10 per cent of the mycorrhizae plated.

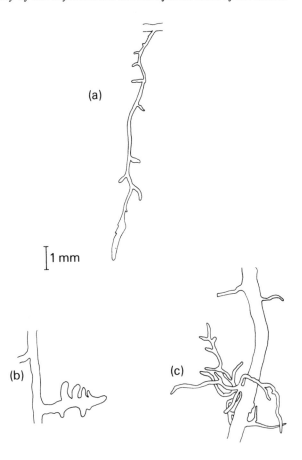

Fig. 18.3. Ectomycorrhizae of *Cotylelobium scarbriusculum* and *Hopea jucunda*. (a) Monopodial type of *Cotylelobium scarbriusculum*. (b) Monopodial type of *Hopea jucunda*. (c) Coralloid type of *Hopea jucunda*.

Two isolates each were obtained from the ectomycorrhizae of *Dipterocarpus zeylanicus* and *Shorea affinis*. Only one isolate each was obtained from *Cotylelobium* and *Hopea. Dipterocarpus hispidus* did not yield any isolates. One isolate from *Dipterocarpus zeylanicus* was initially a white mycelium which turned greenish black within 2–3 days and which became darker as the culture aged. The second isolate from *Dipterocarpus zeylanicus* had a creamish mycelium. Of the two isolates from *Shorea* one had a white mycelium while the other had a blackish mycelium. The isolates from *Hopea* and *Cotylelobium* had white mycelia. Clamp connections were not visible in any of the isolates. The hyphal features of these isolates are shown in Figs. 18.4, 18.5, and 18.6.

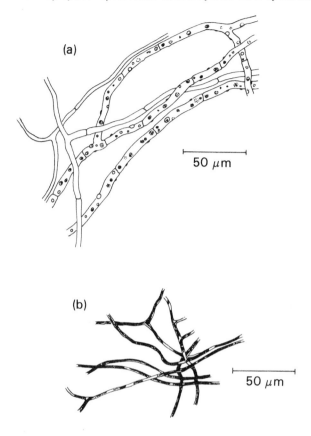

Fig. 18.4. (a) Mycelium of the greenish-black isolate from *Dipterocarpus zeylanicus*. (b) Mycelium of the creamish isolate from *Dipterocarpus zeylanicus*.

Attempts were made to identify the fungal isolates by comparison with pure cultures and also by comparison with cultures obtained from the sporophores of *Suillus* species. The cultural characteristics of the black isolate from *Shorea* agreed with those of the pure culture of *Cenococcum graniforme*. However there was no agreement in the growth rates. The creamish isolate of *Dipterocarpus zeylanicus* on comparison with the culture obtained from the pileus of *Suillus* species revealed that the cultural characteristics were similar but that the growth rates were different. These tentative identifications suggest that *Cenococcum graniforme* and a *Suillus* species could be two of the fungal symbionts associated with the ectomycorrhizae of the forest trees examined.

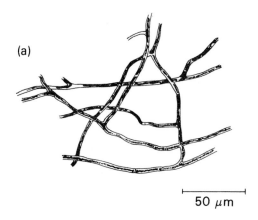

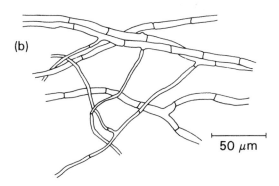

Fig. 18.5. (a) Mycelium of the white isolate of *Shorea affinis*. (b) Mycelium of the blackish isolate of *Shorea affinis*.

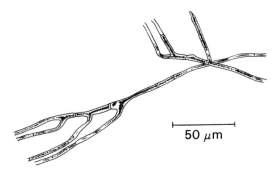

Fig. 18.6. Mycelium of the white isolate of *Cotylelobium scarbriusculum*.

This is in agreement with the work of Muttiah (1970) who showed that the roots of the exotic *Pinus caribaea* var. *bahamensis* of Sri Lanka yielded blackish green *Cenococcum* type fungus from the dry zone and a rusty brown *Boletus* type from the montane zone.

References

Lewis, D. (1961). Genetical analysis of methionine suppressors in *Coprinus*. *Genet. Res.* 2, 141-55.

Meyer, F. H. (1973). Distribution of ectomycorrhizae in native and man-made forests. In *Ectomycorrhizae—their ecology and physiology* (eds. G. C. Marks and T. T. Kozlowski) pp. 79-105. Academic Press, New York.

Mosse, B. (1963). Vesicular-arbuscular mycorrhiza: an extreme form of fungal adaptation. In *Symbiotic associations* (eds. P. S. Nutman and B. Mosse) pp. 146-70. Cambridge University Press.

Muttiah, S. (1970). Initial observations on the introduction of *Pinus caribaea* in Ceylon and certain rooting, transpiration and mycorrhizal studies on seedlings of two provenances of this species, under controlled conditions. *Ceylon Forester* IX, 98-133.

Singh, K. G. (1966). Ectotrophic mycorrhiza in equatorial rain forest. *Malay. Forester* 39, 13-19.

19 Mycorrhizal studies of Iranian forest trees

N. Saleh-Rastin and K. Djavanshir

Abstract

Our preliminary investigation in the Caspian forest of Iran on pine afforestation in humid and arid zones resulted in the determination of ectomycorrhizal association on the following species: *Alnus subcordata, Alnus glutinosa* var. *barbata, Carpinus betulus, Cupressus horizontalis, Fagus orientalis, Parrotia persica, Pinus eldarica, Pinus sylvestris, Quercus castaneifolia, Quercus iberica, Rhamnus grandifolia, Tilia begonifolia,* and *Vaccinium arctostaphyllos.* Ectomycorrhiza were not observed on the following local species when examined: *Acer* spp, *Albizia julibrissin, Buxus hyrcanus, Evonymus* spp, *Fraxinus* spp, *Gleditsia caspica,* and *Pterocarya fraxinifolia.*

The following fungal species and host associations were identified by basidiocarp collection.

Coprinus picaceus on *Quercus iberica.*
Hydnum repandum on *Fagus orientalis* and *Tilia begonifolia.*
Suillus edulis on *Fagus orientalis.*
Lactarius piperatus on *Fagus orientalis.*
Paxillus involutus on *Pinus sylvestris.*
Inocybe dulcamara on *Pinus eldarica.*
Lepiota cristata on *Pinus eldarica.*

Pure culture of these fungi produced enough inoculum for further studies. *Inocybe dulcamara* was thought to be the most important ectomycorrhizal association on the eldar pine *Pinus eldarica* and probably accounts for its successful establishment in the more arid zones of Iran. One of the main plantations of eldar pine is located between Tehran and Karadj on the southern slope of the Alborz range and has the following characteristics. Climate: annual rainfall 255 mm; average maximum temperature of the warmest month 35.7 °C; average minimum of the coldest month –0.5 °C; absolute minimum temperature –15 °C. Soil: sandy texture with a large fraction of small stones which is formed from alluvial limestone; lime represents between 10 and 20 per cent of the soil; pH between 7.5 and 7.9; less than 0.5 per cent organic matter. Natural vegetation: desertic woody shrubs or bushes.

As the annual precipitation is less than 300 mm and there is no rainfall from May to October, the plantation was irrigated every

15 to 20 days during the summer months. A small piece of plantation containing about 180 trees was not irrigated. The trees in this non-irrigated plot are healthy, possibly because it lies in a small depression. Their growth is, however, somewhat less than those under irrigation. In some parts of the area, a few plantings of other trees such as *Ulmus carpinifolia, Robinia pseudoacacia, Fraxinus syriaca, Ailantus glandulosa, Populus alba, Eleagnus angustifolia*, etc. were made but they generally failed, whereas eldar pine prospered.

In this plantation (about 3000 ha) 20 trees in different parts of the area were chosen for mycorrhizal assessments. Soil was removed from around the trunk of trees to a depth of 20 to 30 cm. All the root systems had abundant ectomycorrhizae even though there were no forests on this site before this pine plantation was established 20 years earlier. The seedlings planted in this area originated from around Tehran which has similar ecological conditions and no history of introduction of ectomycorrhizal fungi. Spores of ectomycorrhizal fungi may have been carried into this plantation by water or wind currents. Seedlings in the early plantings had a high mortality rate, but this has improved now that trees are planted in this mycorrhiza infested soil.

In our preliminary investigation the fungi were not identified as no fruiting bodies appeared on the site, but after irrigation fruit bodies of *Inocybe dulcamara* were found. Pure culture of this fungi produced enough inoculum for more research.

20 Vesicular-arbuscular mycorrhiza in Cuba

R. A. Herrera and R. L. Ferrer

Studies on vesicular-arbuscular (VA) mycorrhizae in Central or South America began around 1972 when Mosse (1972; Mosse, Hayman, and Arnold 1973) working with soils and host plants from Brazil, and later Janos (1975), studied the effects of indigenous VA fungi on growth of forest trees in Costa Rica. In Cuba there was no investigation of the Endogonaceae or VA symbiosis before 1973. Only Kreisel (1971) mentioned the occurrence of the genus *Endogone* and pointed out that it produces ectotrophic mycorrhizae in the forests of the island. After 1973 research on VA symbiosis in Cuba has centred on three different aspects.

Occurrence of endogonaceous fungi in natural or man-made vegetation ecosystems

As a result of collecting soil samples in five areas of Western Cuba (a garden sod, three originally forest grasslands, and a natural forest stand) 18 endogonaceous spore types were found and named following a particular nomenclature system—spore Types C (CUBA). This system was established mainly because of the poor accuracy of World keys for Endogonaceae when applied in tropical countries.

Spore types were separated from soil by wetsieving and decanting (Gerdemann and Nicolson 1963). Modified Farrant's medium (arabic gum, 40 g; water, 40 ml; glycerol, 20 ml; and a few crystals of phenol (Davis 1909)) was used for mounting the slides. In five years slides have not shown any change in spore size or wall thickness, but changes do occur in spore content and occasionally in colour.

Fungal types were morphologically described, and comparisons of Types C with those given by Gerdemann and Trappe (1974) were made. Nine spore Types C were comparable to *Glomus mosseae* (Type C_1), *Glomus microcarpus* (Type C_2), *G. fasciculatus* (Type C_3, spores sampled free or in flattened or irregular loose agregations; and Type C_4, spores with outer membraneous wall, sampled free or in clusters), *G. macrocarpus* (Type C_5), *G. caledonius* (Type C_6, though one hyaline and reticulate inner layer was observed), *Gigaspora calospora* (Type C_{10}), *Gigaspora heterogama* (Type C_{11}), and *Sclerocystis coremioides* (Type C_{15}).

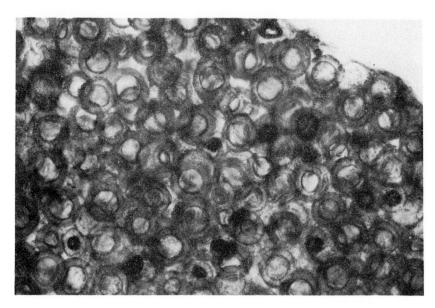

Fig. 20.1. Spore type C_{13}. Sporocarps composed by a mass of spores covered with a thin peridium of anastomosed hyphae.

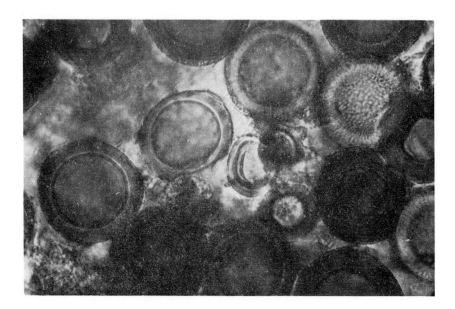

Fig. 20.2. Spores from Type C_{13} showing the inner membraneous and outer thick walls. Note the anastomosed or free spines on the surface of outer wall.

Fig. 20.3. Spore belonging to Type C_{14}. At left the medial (brown) and outer (dark brown) spore walls. At right the inner membrane showing a polar plate with wrinkles.

Fig. 20.4. Sporocarp belonging to Type C_{16}. Probably *Sclerocystis*. Note the thickened walls in the apical regions of spores.

Other spore types collected are probably new species. They seem to be related to the genera *Glomus, Gigaspora,* and *Sclerocystis.*

In many areas of the world it is usual to sample soils with one or two spore types (maximum being five when the soil is cultivated). Mosse and Bowen (1968) have found that 74 per cent of soil samples collected had one spore type present, 20 per cent two, and 6 per cent three. In Cuba we have found 5–11 spore types per soil sample (see Table 20.1) even in sites where there has been little agricultural activity (Sierra del Rosario).

Three characteristic features of Cuban endogonaceous fungi must be pointed out. First of all the floristic richness; even though many mycorrhizal members have not yet been studied, we can establish that the number of species of Endogonaceae per soil sample is to date higher than in any other country in the world. Secondly, the large sporocarps (up to 8 × 5 × 5 mm) as described for *Glomus* by Gerdemann and Trappe (1974) and others have not been found. For example, the largest sporocarp we have found belonging to *Glomus fasciculatus* measured 0.842 mm in diameter. Similarly the mats containing large numbers of sporocarps of *Sclerocystis* have not been observed in Cuba, though *Sclerocystis* is a tropical genus.

Finally, a remarkably feature of the Endogonaceae in Cuba is that in all the soils studied so far, indigenous fungi seem to be affected by tropical conditions: there is a high proportion of empty or dead spores and sporocarps in relation to the living ones. Spores or sporocarps included in empty seed covers were in general alive.

Development of a method to determine the infection density of VA mycorrhizal roots

Phillips and Hayman (1970) designed a valuable method of staining VA mycorrhizal roots. The stain (Trypan Blue) easily penetrates the root tissue and stains hyphae and other components of internal mycelia; the central cylinder may sometimes also be stained, but less deeply.

It was found that the Trypan Blue fixed by fungal components, taking into account no extraction or partial extraction of fixed stain by the vascular bundle, could be extracted. A mixture of lactophenol and 2N HCl (1:1, v/v) extracted the Trypan Blue from the stained internal mycelia after processing stained mycorrhizae by the standard method of Phillips and Hayman (1970) with shaking for one hour at 70 °C.

The spectral analysis of Trypan Blue and the extracts showed that the molecules of stain lose their characteristic absorption in the u.v. range (200–400 nm) after extraction but not in the visible range

Table 20.1. The distribution of the eighteen fungal types in five areas of Western Cuba

Area	Types C (x = presence)																		
	C_1	C_2	C_3	C_4	C_5	C_6	C_7	C_8	C_9	C_{10}	C_{11}	C_{12}	C_{13}	C_{14}	C_{15}	C_{16}	C_{17}	C_{18}	
I	—	—	x	—	—	—	—	—	—	—	x	x	—	—	x	—	—	x	Probably *Glomus*: C_1 to C_9, C_{12} (?), and C_{13} (?)
II	—	x	x	x	x	—	—	—	x	—	—	—	—	—	x	x	—	x	
III	x	—	x	x	—	—	—	—	—	—	—	—	—	x	x	x	x	—	Probably *Gigaspora*: C_{10}, C_{11}, and C_{14} (?)
IV	—	x	x	x	—	x	x	x	—	x	x	—	x	x	—	—	—	—	
V	—	—	x	x	—	—	—	—	—	—	x	x	—	x	x	—	—	—	Probably *Sclerocystis*: C_{15} to C_{18}

I : Managua, La Habana. Ferralitic soil (clay), lightly acid. At present a pasture crop fertilized and irrigated. Originally a forest vegetation.
II : Garden sod from the Institute de Botánica. Clay soil, very much transformed by gardening, high pH.
III : Barrancones, near the Ecological Station of Sierra del Rosario in Pinar del Rio. Clay soil, lightly acid, originally 'yayal' vegetation. At present a natural pasture.
IV : Macurijes, Pinar del Rio. Sandy soil, acid (about pH 5.5). Originally pine-grove vegetation.
V : Ecological Station of Sierra del Rosario, a protected area of seasonal evergreen tropical submontain forest, 'yayal'. Clay soil, lightly acid (about pH 6.5).

(400–700 nm), where the maximum absorption occurs at 590–610 nm as for pure Trypan Blue.

These results suggest that it would be possible to use this method to determine the density of infection of VA mycorrhizae if a curve of concentrations of Trypan Blue is available so that respective optical densities are known.

A more complete study of the method will be carried out in future. The influence of cohabitant fungi, the endophyte anatomy, and phosphate uptake on the Trypan Blue extracted from fungal materials or central cylinder will be analysed in the future. In addition to this, comparisons will be made with the present methods for the determination of VA infection density (Becker and Gerdemann 1977; Hepper 1977).

Occurrence of VA endophytes in forest trees and savannah grasses

In order to determine the presence of VA endophytes of forest trees and savannah grasses, samples were obtained from nurseries of the Centro de Investigaciones Forestales, or obtained by sowing forest tree seeds in polyethylene bags with soil, or sampled directly from the field.

Table 20.2. VA mycorrhizal plant species in Cuba

Name	Local name	Family
Forest species		
Calophyllum antillanum var *brasiliensis*	Ocuje	*Clusiaceae*
Cedrela odorata	Cedro	*Meliaceae*
Melia azedarach	Paraiso	*Meliaceae*
Swietenia macrophylla	Caoba de Honduras	*Meliaceae*
S. mahagoni × *S. macrophylla*	Caoba Hibrida	*Meliaceae*
Cordia gerascanthus	Baria	*Boraginaceae*
C. alliodora	Baria Amarilla	*Boraginaceae*
Hibiscus elatus	Majagua	*Malvaceae*
Spathodea campanulata	Tulipán	*Bignoniaceae*
Tabebuia angustata	Roble Blanco	*Bignoniaceae*
Pithecellobium dulce	Inga Dulce	*Leguminosae*
Savannah species		
Paspalum notatum	Bahia	*Gramineae*
Panicum maximum	Guinea	*Gramineae*
Cynodon dactylon	Bermuda	*Gramineae*
Digitaria sp		*Gramineae*
Cassia bifila		*Leguminosae*
Centrosema pubescens		*Leguminosae*

Roots from different species were cut in segments and stained by the method of Phillips and Hayman (1970). Only the presence or absence of VA endophytes was recorded. All species were observed to be mycorrhizal (Table 20.2).

References

Becker, W. N. and Gerdemann, J. W. (1977). Colorimetric quantification of vesicular-arbuscular mycorrhizal infection in onion. *New Phytol.* 78, 289-95.

Davis, W. B. (1909). Farrant's medium for mounting mosses. *Bryologist* 12, 8.

Gerdemann, J. W. and Nicolson, T. H. (1963). Spores of mycorrhizal *Endogone* species extracted from soil by wet sieving and decanting. *Trans. Br. mycol. Soc.* 46, 235-44.

— and Trappe, J. M. (1974). The Endogonaceae in the Pacific Northwest. *Mycologia Memoir* No. 5.

Hepper, C. M. (1977). A colorimetric method for estimating vesicular-arbuscular mycorrhizal infections in roots. *Soil Biol. Biochem.* 9, 16-18.

Janos, D. P. (1975). Effects of vesicular-arbuscular mycorrhizae on lowland tropical rainforest trees. In *Endomycorrhizas* (eds. F. E. Sanders, B. Mosse, and P. B. Tinker). Academic Press, London.

Kreisel, H. (1971). Clave para la identificación de los macromicetos de Cuba. *Ciencias* (La Habana), Serie 4, Ciencias Biológicas, No. 16, p. 3.

Mosse, B. (1972). Effects of different *Endogone* strains on the growth of *Paspalum notatum. Nature, Lond.* 239, 221-3.

— and Bowen, G. D. (1968). The distribution of *Endogone* spores in some Australian and New Zealand soils and in an experimental field soil at Rothamsted. *Trans. Br. mycol. Soc.* 51, 485-92.

— Hayman, D. S., and Arnold, D. J. (1973). Plant growth responses to vesicular-arbuscular mycorrhiza. V. Phosphate uptake by three plant species from P-deficient soils labelled with [32]P. *New Phytol.* 72, 809-15.

Phillips, J. M. and Hayman, D. S. (1970). Improved procedures for clearing roots and staining parasitic and vesicular-arbuscular mycorrhizal fungi for rapid assessment of infection. *Trans. Br. mycol. Soc.* 55, 158-61.

PART IV

Mycorrhizal nutrition of tropical plants

21 Mycorrhizal roles in tropical plants and ecosystems

G. D. Bowen

Introduction

It has been amply demonstrated that inoculation of forest trees and agricultural plants with mycorrhizal fungi can stimulate their growth in nutritionally poor soils, such as occur in very large areas of the tropics. The possible impact of mycorrhizae can be seen from the fact that almost all plant species of economic importance in the tropics are able to be infected; most form vesicular-arbuscular (VA) mycorrhizae but some tree species form ectomycorrhizae (Gerdemann 1975; Harley 1969). The only notable exception is rice growing under padi conditions, where infection is rare.

A number of mycorrhizal fungi can infect a plant species and, therefore, one may be able to modify nutrient absorption by the plant considerably by judicious selection of a fungus. Such manipulation may be valuable in management of nutrient deficient soils and in plant breeding programmes which aim at selecting plant varieties tolerant to deficiencies. I believe also that mycorrhizal fungi are key components of the efficient, closed, nutrient cycle of natural systems and have effects on the expression of plant disease. The plant scientist has considerable scope to manage plant production imaginatively as he is dealing with a dynamic three-member system, viz. soil, plant, and mycorrhizal fungi rather than with only soil and plant.

A selected efficient fungus must be able to be introduced into existing soils in the face of competition from nautrally occurring micro-organisms. Some facets of this most important topic have been reviewed by Bowen (1978) and will be raised in later chapters. Here I deal more with modes of mycorrhiza function; hopefully an understanding of these will help us see the advantages and limitations of mycorrhizal associations. I shall deal only with ectomycorrhizae and vesicular-arbuscular mycorrhizae; they will be treated together for much the same considerations apply to each. Mycorrhizae in orchid culture, an important industry in some tropical countries, has been reviewed by Smith (1974). Detailed discussion of the biology of mycorrhizae is given in Harley (1969), Marks and Kozlowski (1973), and Sanders, Mosse, and Tinker (1975).

In this paper I shall be trying to relate what is known of mycorrhizal function (almost all studies of function have been on temperate climate species) to plant growth in tropical soils and to tropical systems of production. However, there is no one 'tropical' system or environment, for a wide range of soils, climates, and land uses occur in the tropics. For example approximately half of the tropics has a marked wet and a dry season, a quarter has a high evenly distributed rainfall and a quarter is semi-arid or desert. Each of these lead to different soil problems for plant growth and management. Furthermore, savannah occupies some 43 per cent of the area and tropical rain forests some 30 per cent (Sanchez 1976), each special in its own way; 45 per cent of the agricultural activity is as shifting cultivation, about another 19 per cent is farming (often at a subsistence level), and 4 per cent is under plantation. One factor common to most of these situations is the need to increase and *sustain* productivity (with low fertilizer imputs if possible). In many cases this means an increasing emphasis on intensified farming and a moving away from shifting cultivation. In other cases it may mean optimizing productivity in the shifting cultivation system and then optimizing site recovery.

Many of the tropical agricultural systems are unique and particularly involve plants growing together rather than in monocultures which dominate temperate climate farming. A further unique aspect of tropical soils is their high temperatures, e.g. at Ibadan (Nigeria) soil temperatures can be 42 °C at 5 cm depth (Sanchez 1976). When a forest soil is cleared, surface soil temperatures can increase by up to 11 °C, and this is accompanied by an increased rate of organic matter decomposition, leaching of nutrients, and potential degradation of soil resources. Mycorrhizal management may have a role in ameliorating this decline and enhancing crop production. The major fungal species causing VA mycorrhiza are the same in tropical countries as in other parts of the world (e.g. Redhead 1977) but other types of mycorrhizal fungi may be found when more study is made in the tropics. In any case, considerable ecotypic variation occurs within ectomycorrhizal and endomycorrhizal fungi in a very wide range of properties (Trappe 1977; Theodorou and Bowen 1971; B. A. Daniels, personal communication). It may be necessary to select fungal strains, within a species, which perform well in particular environments, e.g. high soil temperatures. There is also evidence (e.g. Moawad 1979) that differences occur between mycorrhizal and non-mycorrhizal plants in their response to high soil temperature and therefore conclusions from studies performed on non-mycorrhizal plants may have to be modified considerably for mycorrhizal plants.

Nutritional responses to mycorrhizae

The literature has many reports of responses of several plant species (many of them tropical) to inoculation with mycorrhizal fungi (Mosse 1979). These are often in sterilized soils in the glasshouse but some reports exist for field studies, where responses are still striking but sometimes diminished because of naturally occurring mycorrhizal fungi and/or difficulty in establishing inoculum. The great importance of ectomycorrhizal inoculation of introduced pine species in the tropics has been reviewed by Mikola (1970).

Some typical examples of mycorrhizal responses are given in Table 21.1. Large increases in root and shoot growth and in the uptake of a number of ions occur from inoculation, due partly to the increased

Table 21.1. Plant responses to mycorrhizal inoculation

(a) Ectomycorrhizae on *Pinus elliottii* var. *elliottii* (Lamb and Richards 1971). Plants were grown in pots in a peat–sand mix for 8 months

Inoculum	Dry weight (g)	Nutrient content (mg)		
		N	P	K
Nil	0.26	2.44	0.22	2.27
Isolate R. 8.35[+]	1.11	11.14	1.12	10.72
Rhizopogon sp	1.96	25.15	2.20	23.77
Isolate R. 8.22[+]	2.81	38.92	3.86	35.60

(b) Ectomycorrhizae on *Pinus caribaea* (Vozzo and Hacskaylo 1971). A field study in Puerto Rico—age, 40 weeks

Inoculum	Height (cm)
Nil	8.8
Rhizopogon roseolus	11.4
Corticium bicolor	11.4
Suillus cothurnatus	13.3

(c) VA mycorrhizae on *Araucaria cunninghamii* (Bevege 1971). Plants were grown for 26 months in a fertile manured red loam (total P 1600 p.p.m.).

Plant variable	Inoculated	Uninoculated
Dry wt. (g)	76.1	7.2
Shoot N%	1.25	1.88
Root N%	0.96	0.94
Shoot P%	0.079	0.042
Root P%	0.109	0.053

Table 21.1 *(cont.)*

(d) VA mycorrhizae on *Paspalum notatum* (Mosse 1972). Plants were grown for 11 weeks in an irradiated-sterilized Brazilian soil (2.9 p.p.m. sodium bicarbonate soluble P)

Inoculum*	+ lime (pH 5.8)	– lime (pH 4.8)
E_3†	115	58
Honey-coloured†	41	65
Indigenous strain	79	14
Laminate†	33	16
Nil	10	11

*Inocula are designated after Mosse and Bowen (1968). *Trans. Br. mycol. Soc.* 51, 469–83.
†E_3 and honey coloured are probably synonymous with *Glomus fasciculatus* and *Acaulospora laevis* respectively. (Gerdemann and Trappe (1974). *Mycol. Mem.* 5.)

absorption of elements in short supply (the mycorrhizal response) and partly to increased uptake of other ions which then occurs with increased plant growth. In many cases, e.g. grasslands, the full response for an initially limiting nutrient such as phosphate is not obtained because a second nutrient, e.g. nitrogen, may become limiting (Koucheki and Read 1976).

Figure 21.1 indicates the way in which a mycorrhizal response should be assessed, viz. uninoculated plants with nutrients added (in this case phosphate) were compared with inoculated plants at various phosphate applications. Note: (i) differences occur between mycorrhizal fungi in stimulating plant growth, (ii) the mycorrhizal response was eliminated by the addition of phosphate thus indicating the response was involved with phosphate nutrition, and (iii) by matching plant growth in inoculated and uninoculated plants at different phosphate applications, the phosphate equivalence of the response, and the fertilizer saving, can be calculated. With perennials, e.g. pasture species, it is appropriate to measure not only the plant's establishment requirement for nutrients but also the maintenance requirements after establishment (Kerridge 1978) and this concept should also apply to measurement of mycorrhizal benefits.

In studies of the mechanism of response, the growth of mycorrhizal and non-mycorrhizal plants of the same nutrient content should also be compared. If growth is similar, as is often the case, one can conclude that the main mechanism of response to mycorrhizae was increased efficiency of nutrient uptake. However in some cases (e.g. Bowen 1973; Abbott and Robson 1977) they are different and this

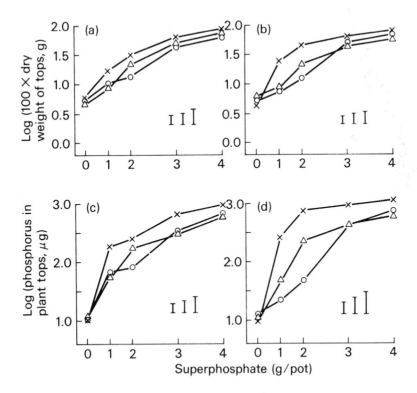

Fig. 21.1. Effect of VA mycorrhiza and phosphate applications on growth and phosphorus concentration of *Trifolium subterraneum* after four weeks. (a) and (c) are natural soil, (b) and (d) are the same soil steamed to remove natural mycorrhizal fungi. X and △ = different VA fungi, ○ = control. Vertical bars represent LSDs at *P* = 0.05, 0.01, and 0.001 respectively. (From Abbott and Robson 1977, by permission *Australian Journal of agricultural Research*).

suggests factors other than uptake are also operating. Abbott and Robson found in steamed soil, but not in unsterilized soil, non-mycorrhizal subterranean clover produced more dry matter at a given phosphorus concentration in tops than did mycorrhizal. This aspect of mycorrhizal response deserves more study although it is well accepted now that the *main* mycorrhiza response is one of increased efficiency of nutrient uptake.

In Fig. 21.1 the response was mainly a phosphate response but increased absorption of potassium, sulphate, zinc, and strontium-90 (a calcium tracer) by mycorrhizal plants has also been shown experimentally (Powell 1975; Gray and Gerdemann 1973; Gilmore 1971;

Jackson, Miller, and Franklin 1973). The increase in uptake of potassium (a moderately mobile ion in soil) shown by Powell (1975), some 23 per cent, was not as dramatic as many recorded increases in absorption of phosphate (a poorly mobile ion). Differences have been shown sometimes between mycorrhizal and non-mycorrhizal plants in their concentration of nitrogen, calcium, sodium, magnesium, iron, manganese, copper, boron, and aluminium (Mosse 1973a). On general grounds (see below) one might expect increases in uptake of the more poorly mobile ions such as molybdenum, iron, copper, cobalt, and sometimes manganese.

A number of conclusions come from the mycorrhizal stimulation of nutrient uptake and growth such as in Table 21.1 and Fig. 21.1:

  (i) Infection with an appropriate mycorrhizal fungus can radically change the estimate of production potential of a soil and its fertilizer requirement.

 (ii) Nutritional studies on tropical species, e.g. ranking their relative fertilizer requirements, can be badly in error by neglecting the mycorrhizal factor. Almost no nutritional studies on tropical or temperate plants have ensured adequate mycorrhizal infection was present.

(iii) The search for plant genotypes capable of tolerating deficiency conditions, now being carried out by many agricultural scientists must recognize the importance of using appropriate mycorrhizal fungi in test programmes. These fungi may short-circuit and facilitate the selection-breeding programmes. Selection for productivity under deficiency conditions involves selection for efficient uptake of nutrients in low supply and for efficient use of absorbed nutrients. It may be that the former of these can be achieved by endomycorrhizal fungal selection.

(iv) The endomycorrhizal increase in ion uptake can have large consequences for nitrogen fixation by legumes and by non-legumes such as *Casuarina* spp which are endomycorrhizal (El-Giahmi, Nicolson, and Daft 1976). VA mycorrhizae have been demonstrated to increase uptake of phosphate, sulphate, and zinc and probably would increase uptake of cobalt, molybdenum, copper, and iron, all of which are involved in nitrogen fixation. Mosse, Powell, and Hayman (1976) found clover, *Centrosema*, and *Stylosanthes* did not nodulate in grossly phosphate-deficient soil if they were not also mycorrhizal.

Modes of function of mycorrhizae in nutrition

All too often the physiology of a large mycorrhizal plant is wrongly compared with that of a smaller, nutrient deficient non-mycorrhizal plant. This is quite wrong; the correct procedure is to compare the mycorrhizal plant with a fertilized non-mycorrhizal plant of the same size and/or nutrient content, in order to distinguish unique mycorrhizal effects on physiology, unrelated to the enhanced nutrient uptake they cause. The enhanced nutritional status in mycorrhizal plants was probably the basic reason for changes in stem anatomy and for enhanced virus production recorded by Daft and Okusanya (1973a,b), as they themselves pointed out. Examination of the lower root:shoot ratios of mycorrhizal plants shows these are often similar to those of non-mycorrhizal plants of the same nutrient status, and not an effect peculiar to the mycorrhizal infection.

Mechanisms of increased nutrient absorption by mycorrhizae

Principles of ion uptake from soil, especially in relation to ecto-mycorrhizae have been discussed by Bowen (1973). The same principles apply to VA mycorrhizae. In brief, ion uptake is governed by two factors: the absorbing capacity of the root (the property usually measured in uptake experiments in stirred solutions) and the transfer of ions to the root through soil. For highly mobile ions in soil, e.g. nitrate and to some extent sulphate and potassium, differences between parts of roots or between species of plants in absorbing capacity will determine their relative uptakes of these nutrients. In a monoculture, relatively low concentrations of roots through the profile will be adequate for uptake of available supplies of such ions in soil (Barley 1970). However, for poorly mobile ions, such as phosphate, zinc, copper, molybdenum, and to some extent potassium and ammonium ions, the rate-limiting factor is their movement to the root. In these cases root length and morphological modifications of the root, such as root hair development and growth of mycorrhizal fungi into soil, largely determine ion uptake.

Compared with adjacent uninfected tissue or non-mycorrhiza laterals, VA mycorrhizae and ectomycorrhizae have been shown to have increased absorption of phosphate, sulphate, and zinc from stirred solutions (Morrison 1962; Harley 1969; Bowen 1973; Bowen, Skinner, and Bevege 1974). The influx is often similar to that of rapidly expanding apical portions of uninfected roots (Bowen 1973) but these do not maintain their high activity for as long as mycorrhizae. For the reasons stated above increased absorbing capacity is likely to be of little ecological importance for uptake of phosphate and zinc except possibly in situations approaching solution uptake, e.g. material in leaf leachates before it is absorbed by soil or litter, or local high concentrations of these ions.

Large differences occur between species in their production of roots and root hairs and these are important in uptake of non-mobile and poorly mobile ions. At one end of the spectrum many tree species have low root concentrations and few or no root hairs; these have been designated as 'magnolioid' roots (Baylis 1975). With such plants only a relatively small part of the poorly mobile ions in soil are used by the roots. 'Graminoid' roots are at the other end of the spectrum and these have high root lengths per cm^3 of soil, mainly fine roots and usually with well developed root hairs. In such cases the soil is used very effectively and often the feeding zones of adjacent roots soon overlap. In acute deficiency situations, however, root concentrations even of these decline markedly due to lack of assimilate from the above ground parts (Bowen, Cartwright, and Mooney 1974) and again the soil volume would not be used fully. A range of root morphology occurs between the extremes of magnolioid and graminoid roots, e.g. in many tropical legumes such as *Arachis* and *Stylosanthes* few root hairs are formed, and these are mainly in clumps at the points of emergence of lateral roots. Furthermore, even in plant species with the ability to form many root hairs, these are often produced only sparsely because many factors can affect their development, e.g. soil nutrient and moisture levels, soil physical factors and certain micro-organisms (Cormack 1962; Bowen and Rovira 1961; Reid and Bowen 1979). Table 21.2 (from Barley 1970) indicates the variation in rooting intensity between species and depth. There is

Table 21.2. Root lengths per cm^3 of soil (L_v) for plants in the field (from Barley 1970)

Species	Depth (cm)	L_v (cm/cm^3)
Herbs: Gramineae		
Poa pratensis	0–15	50
Cereals (oats, rye, wheat)	0–15	5–25
	25–50	4
Herbs: Non-Gramineae		
Stylosanthes gracilis	0–10	30
	40–50	3
Medicago sativa	0–10	20
Glycine max	0–15	4
Woody plants		
Tea (*Camellia sinensis*)	0–2.5	4
	45–47.5	1
Pinus radiata	0–8	2
	25–45	0.8

relatively little quantitative information on rooting in tropical plants and crops.

It has been suggested on theoretical grounds and experimentally demonstrated for ectomycorrhizae and endomycorrhizae that the main mode of increased uptake of poorly mobile ions is by fungal growth into soil, absorption of ions, and their transport back to the higher plant (Bowen 1968, 1973; Skinner and Bowen 1974a; Sanders and Tinker 1973; Hattingh, Gray, and Gerdemann 1973). Sanders and Tinker (1973) found with endomycorrhizal onions that there could be 80 cm of hyphae/cm of infected root. Greatly increased longevity of the mycorrhiza (up to several months for ectomycorrhizae) in the one position in soil can also increase radii of diffusion of ions to the root (Bowen 1968) but this is likely to be of less general importance than the penetration of soil and litter by mycelia and will be important particularly in sandy soils with low buffering power and moderate levels of phosphate. Sanders, Tinker, Black, and Palmerley (1977) found the extra inflow of phosphate into mycorrhizal onion roots was closely related to the percentage of root infected by the fungi which in turn was correlated with production of external hyphae of the VA mycorrhiza. It is at this position that epidemiological factors such as population of fungal propagules in soil and spread of infection in the plant interact with plant response (see Bowen and Rovira 1976).

Because of their poor rooting intensity and poor production of root hairs, magnolioid roots respond greatest to mycorrhizal infection, e.g. *Araucaria cunninghamii* (Table 21.1), and the relative response of various species tends to decline as roots become more graminoid (Baylis 1978). Table 21.3 from Crush (1974) on legumes, illustrates well the much greater responses to mycorrhizal infection when root hairs are poorly developed.

A vast number of tropical soils are sufficiently low in available phosphate for even grasses to respond markedly to mycorrhizal inoculation. Unfortunately, available phosphate in soil is often not recorded in experiments, but in pot trials in an autoclaved Nigerian soil with 15.5 p.p.m. available phosphate (relatively high for tropical soils), Sanni (1976) increased grain yield of rice (under upland conditions) by 40 per cent; in a Brazilian soil with 2.4 p.p.m. sodium bicarbonate soluble phosphorus, Mosse, Hayman, and Arnold (1973) increased growth of *Paspalum notatum* in pots seven fold by inoculation with appropriate VA fungi. In the field in Pakistan, mycorrhizal inoculation of maize increased growth by 50 per cent (Khan 1972). Even where surface soil concentrations of grass roots are high, they are frequently low enough in sub-surface horizons (Table 21.2) for us to expect mycorrhizae to have a supplementary nutritional

Table 21.3. Root response to VA infection (from Crush 1974)

Factor	Plant Centrosema*	Stylosanthes*	Trifolium*	Lotus*
Weight of tops				
(g FW/plant)†				
Control	1.67	0.47	1.56	2.54
Mycorrhizal‡	3.88	1.63	2.57	2.01
Added phosphate				
(not inoculated) §	4.95	0.91	3.97	3.86
Root characters				
Root diameter (μm)	288	285	221	229
Root hair length (μm)	106	108	213	809
Roots with root				
hairs (%)	17	6	79	99

*All plants were inoculated with *Rhizobium.*
†Harvested at seven weeks.
‡Inoculated with endophyte E₃ (≡ *Glomus fasciculatus*).
§ 0.4 g monocalcium phosphate/kg soil.

function by acting there. Mycorrhizal infection can occur to 50 cm depth although sometimes at a much reduced level (Sutton 1973; Bowen, unpublished).

The pattern of uptake of poorly mobile ions by plants is, we can now conclude, one first of intensive depletion of a narrow zone around the root, delineated by the root hairs if they occur, and limited depletion around them depending on the ion and the soil (Bhat, Nye, and Baldwin 1976), and then with mycorrhizal infection, soil further out is explored, sometimes quite extensively.

Many experiments indicate that mycorrhizal plants use the same sources of inorganic phosphate in soil as do plant roots (Hayman and Mosse 1972; Barrow, Malajczuk, and Shaw 1977) but have greater access to them by growth of the fungus hyphae supplementing the plant roots. This also appears to be the way in which mycorrhizal responses to poorly soluble phosphate such as rock phosphate occur. Access to poorly soluble forms of phosphate is of major interest in some tropical soils especially where fixation of phosphate by iron and aluminium occurs but use of these poorly soluble phosphates seems to be small. Sanni (1976) did not find mycorrhizal rice able to use strengite, a major form of insoluble phosphate in Nigerian soils. For some time it has been hypothesized that oxalates and possibly other compounds excreted by fungi, particularly ectomycorrhizal basidiomycetes, may solubilize phosphate or chelate other compounds in soil. Although the potential for this exists, experimental evidence for its being a major part of the mycorrhizal response is still lacking.

Of particular interest is the finding of Azcon, Barea, and Hayman (1976) that lavender inoculated both with a mycorrhizal fungus and phosphate-dissolving bacteria in the presence of rock phosphate, absorbed more phosphate than plants with either mycorrhizae or bacteria alone. The mechanism of this synergistic effect is not certain but needs further study as does the generality of the response, for it may not occur in all cases (see Barea, Azcon, and Hayman 1975).

The use of organic phosphate by mycorrhizae has received rather less attention than use of inorganic phosphates but is pertinent to recycling of nutrients in litter in both savannah and rain forest ecosystems. Also, in some highly weathered soils of the tropics and in soils of the tropical highlands (Andepts) organic phosphorus accounts for over half of the total soil phosphorus. The demonstration of phytases and various phosphatases associated with ectomycorrhizae and VA mycorrhizae (Theodorou 1968; Bowen 1973; Williamson and Alexander 1975; R. C. Foster, unpublished) raises a good case for mycorrhizae aiding the plant in uptake of organic phosphates. Phosphatase activity occurs with uninfected plant roots growing in sterile conditions and it is probable that the uptake of phosphates from litter occurs not by a unique property of the fungi but by their advantageous spatial positioning (compared with the root) in penetrating litter and organic matter (see below).

The consequences of fungal growth into soils

One of the keys to increased ion uptake by ecto- and VA mycorrhizae is the nutrient absorption by the fungi growing from the roots into soil. Different responses to various ectomycorrhizal and VA fungi may be related to their relative ease of infection, differences in rate of spread through the root system, and consequent relative growth of the hyphae into soil and these factors need more study. Bowen (1973) indicated the more efficient ectomycorrhizal fungi produced mycelial strands readily.

Considering the importance of mycorrhizal fungal growth into soils, surprisingly little experimental study has been performed on factors affecting this. Rhodes and Gerdemann (1975) observed hyphal growth from VA mycorrhizae of onions over and near the surface of soil for 8 cm. In a plantation of *Pinus radiata*, Skinner and Bowen (1974a) demonstrated translocation of phosphate in mycelial strands of ectomycorrhizae for 12 cm. Using an experimental approach to mycelial strand growth into soil, Skinner and Bowen

(1974b) showed up to 100-fold differences in penetration of different soils by mycelial strands. Compacting soil from 1.2 g/cm^3 to 1.6 g/cm^3 reduced mycelial growth by up to 90 per cent and this would be expected to reduce nutrient uptake considerably. Hyphal growth into soil is an important area for future research.

There are many possible roles of mycorrhizal fungal growth from roots into soils, quite apart from that discussed above and apart from their role in spreading the infection to nearby roots. The use of mycorrhizal fungi for root functioning in many potentially deleterious situations is an important alternate strategy evolved by plants. I have briefly listed some of the possible roles as I see them, although in a number of cases experimental testing is needed. Most are particularly relevant to tropical soils.

Uptake of poorly mobile or immobile ions beyond the root hair zone (discussed above).   This is the most general function and obviously the most efficient fungi for particular soils must be selected and ways found to introduce them successfully.

Absorption of water.   Water relations of mycorrhizae have been little studied but may be particularly important in sandy soils and in the semi-arid tropics. Sands and Theodorou (1978) found the hydraulic conductivity of ectomycorrhizae and uninfected roots of *Pinus radiata* to be similar. However, some ectomycorrhizal fungi can grow at much lower water potential than higher plants (Theodorou 1978) and thus may make a significant extra contribution to water uptake by the root in some soils.

The main interactions between mycorrhizae and water are likely to be (i) fungal growth into soils may increase the rate of water movement to the root, especially with low soil moisture in sandy soils and especially with low rooting intensities. In a sandy soil, Sands, Greacen, and Gerard (1979) have calculated that the hydraulic conductivity of soil falls by five to ten orders of magnitude between field capacity (–10 kPa matric potential) and wilting point (–1500 kPa) and in such soils this transport factor often becomes the rate limiting factor in water uptake, (ii) fungal growth into soil may give a greater opportunity for 'root'-soil contact, compensating for any shrinkage of roots from soil as it dries, and (iii) probably most important, mycorrhizae should increase absorption of ions which are relatively highly mobile in saturated soils but poorly mobile in unsaturated soils, e.g. K$^+$. The apparent diffusion coefficient of highly mobile ions can decrease by more than an order of magnitude between field capacity and wilting point (e.g. Rowell, Martin, and Nye 1967).

Counteracting toxicities (e.g. low pH, high aluminium, high salt, high alkalinity). (i) It is possible, but yet untested, that mycorrhizal fungi could increase absorption of elements harmful to plants and animals eating those plants, e.g. cadmium, and certain organic residues. Fortunately these are likely to be only very rare situations. The possible interactions of toxic elements and mycorrhizae have been discussed by Bowen (1978).

(ii) It is important to remember the flexibility built into a plant system by having an alternative type of absorbing organ, i.e. fungal hyphae in the form of mycorrhizae. In highly acid soils of the tropics, the main effects on plants can be by aluminium toxicity, one effect of which is to injure root tips and prevent root growth. Theoretically, fungal growth into soils from mycorrhizae could compensate for this loss of root growth provided they are not also affected. Aluminium tolerant varieties of plants often are tolerant of low phosphate situations (Sanchez 1976)—could there be a mycorrhizal component in this? Some aluminium tolerant species have a lower translocation of aluminium to the tops—could mycorrhizal fungi act similarly by being a sink of aluminium? R. C. Foster (unpublished) has found lead and chloride to be accumulated in some ectomycorrhizal and VA fungi. These factors need further research but the selection of particular fungi for such situations may be possible. Some mycorrhizal fungi are far more suited to acid conditions than others and selection of them for use in acid soils is important.

Similarly, the mycorrhizal situation with highly saline and alkaline soils of the semi-arid tropics needs further study. Mosse (1972) has shown that some mycorrhizal fungi perform very much better than others depending on the soil pH. Khan (1974) found frequent VA infection of halophytes in salty soils but no measurements of salinity were made. Bowen (unpublished) has found high populations of typical VA spores in soils with greater than 5000 p.p.m. Cl⁻ associated with extremely slight rises in salt pans and at least some infection under these extreme conditions. In other soils typical mycorrhizae were found at pH 9.2. The frequent occurrence of VA mycorrhizae in coastal dunes was recorded by Nicolson (1960). Whether mycorrhizal infection substantially assists plant tolerance of high alkalinity or high salinity is a question well worth experimental study.

(iii) An environmental stress relevant to most of the tropics is high soil temperatures as discussed in the introduction. Marx and Bryan (1971) found *Pinus taeda* ectomycorrhizal with *Pisolithus tinctorius* and maintained at 40 °C soil temperatures for five weeks, grew as well as those at 25 °C soil temperature but 55 per cent of the non-mycorrhizal plants died at 40 °C. Non-mycorrhizal plants were usually smaller than mycorrhizal plants, but Marx and Bryan also found the larger non-mycorrhizal plants were more susceptible to high temperatures. More detailed physiological study with nutritionally matched plants is required, but these findings and those of Moawad (1979) with VA mycorrhiza suggest possible physiological differences between mycorrhizal and non-mycorrhizal plants of direct relevance to growth in tropical soils. Changing soil temperatures can markedly affect the development of different ectomycorrhizal and VA fungi (Marx, Bryan, and Davey 1970; Crush 1973) and this may be important in tropical soils where temperatures can rise by 11 °C after clearing (Sanchez 1976). Laboratory studies on the growth of ectomycorrhizal fungi at various temperatures may have little relevance to their growth in association with roots (Theodorou and Bowen 1971).

Structural effects.   Nicolson (1960) found high VA infection of many grasses on developing sand dunes. They probably assisted ion uptake by the grasses but could the fungal growth have assisted in stabilizing the dunes? Recently Sutton and Sheppard (1976) showed by inoculating sterilized dune sand with *Glomus*, pots with endomycorrhizal *Phaseolus vulgaris* had some five times the weight of sand aggregates/kg as pots with non-mycorrhizal plants of similar size, the sand grains being held together by the fungal hyphae. The possible structure effects of VA mycorrhiza warrants further examination in poorly structured soils, especially where intensive agriculture could lead to loss of structure as in some high rainfall areas of the tropics.

Nutrient conservation.   In natural communities in the high rainfall areas of the tropics leaching losses from soils are remarkably small, despite high temperatures and good conditions for decomposition of litter. In a tropical rain forest in Puerto Rico, Odum and his coworkers (1970) found ^{32}P released from decaying leaves was quantitatively captured by surface roots in the top 5 cm of soil and subsequently absorbed by plants; potassium was recycled very quickly (but loosely held) and calcium, magnesium, manganese, iron, and copper were cycled slowly and bound very tightly. Went and Stark (1968) suggested interception of nutrients in leachates by hyphal growth from mycorrhizae was the key to the closed nutrient cycles in tropical rain forests but no experimental validation of this was attempted and it is doubtful if such interception would approach 100 per cent. Furthermore, much of the profuse mycelial growth under forests may be associated with litter decomposers, not mycorrhizal fungi. Haines and Best (1976) purported to demonstrate in soil columns that the presence of VA mycorrhiza on *Liquidambar styraciflua* significantly reduced nitrogen loss by leaching. However, mycorrhizal plants were several times the size of non-mycorrhizal plants and the observed large reduction in leaching of nitrogenous compounds may well have been due to the larger root systems of mycorrhizal plants and not to the mycorrhizae themselves.

In forests, and in grasslands, it is probable that mycorrhizal associations play an important part in the conservation of nutrients and in nutrient cycling. Fungal hyphae readily penetrate litter and decomposing organic matter and can spatially compete with other soil micro-organisms for organic and inorganic nutrients far more efficiently than can plant roots (Bowen 1973). This is probably an important part of the tight nutrient cycle of rain forests and other systems. This spatial-competition factor and feeding in the litter layers is likely to be of far more general importance in nutrition from organic matter than only the production of enzymes such as phosphatases by the fungi. A possible expression of this competition factor is seen in the work of Gadgil and Gadgil (1975) who found litter decomposition of *Pinus radiata* was decreased by the presence of mycorrhizae—they hypothesized absorption of nutrients by the mycorrhizal fungi suppressed decomposition by removing nutrients needed by decomposer micro-organisms. N. Stark (personal communication) has recently demonstrated absorption of ^{32}P in leaf material by mycorrhizal fungi and its translocation to the plant. One does not have to hypothesize that the mycorrhizal hyphae intercept leachates (although they may)—one has only to have nutrients in, or fixed on, the litter and then have mycorrhizal fungi successfully compete with other micro-organisms for these nutrients. The probable key role of mycorrhizal fungi in regulating nutrient flow in natural systems, both forest and savannah demands much further study.

Compensation and avoidance of disease.   See pages 180-1.

Claims for nitrogen fixation by mycorrhizae

Repeated claims for nitrogen fixation by mycorrhizae must be viewed with the greatest caution. In no cases have the mycorrhizae been free of other organisms—significant nitrogen fixation has so far only been recorded with procaryotic systems such as bacteria and blue-green algae. It has also been suggested that the observed nitrogen fixation in laboratory experiments with mycorrhizal plants has been due to fixation by free-living bacteria encouraged by the mycorrhizae but few quantitative or qualitative studies of nitrogen fixing bacteria associated with mycorrhizae and non-mycorrhizal roots have been carried out. Studies on the hyphosphere of mycorrhizae, while often showing increased general microbial activity are inadequate in the present context.

A much more disturbing facet of claims for nitrogen fixation by mycorrhizae or their associated microflora is the quite unecological conditions under which the hypothesis is tested; plants are sometimes exposed in a sealed tube of limited volume under high humidity conditions for several days (e.g. Bevege, Richards, and Moore 1978). This would give good conditions for abnormal development of bacteria, including nitrogen fixers, on roots and foliage. Great difficulty can be experienced in washing bacteria from the rhizosplane and 'root incorporation' of ^{15}N in such experiments may be really in micro-organisms. The tendency not to use similar non-mycorrhizal plants as controls is a further very disturbing facet of many such studies. Giles and Whitehead (1977) have raised the extremely interesting possibility of incorporating protoplasts of nitrogen-fixing bacteria into mycorrhizal fungi and this could be an avenue well worth studying with both ecto- and endomycorrhizae.

Mycorrhizae and disease expression

Root diseases caused by such organisms as *Phytophthora* spp are common in tropical countries and may be reducing yield considerably in poor soils without the disease being obvious. What a plant often 'sees' in root diseases is a loss of roots important for nutrient and water uptake. In a rich soil some loss of roots is of little consequence but in a nutrient-poor soil the same loss of roots may decrease yield appreciably.

Mycorrhizal interaction with root disease can occur in four ways:

Antibiosis.   Marx (1973) has indicated that some ectomycorrhizae can produce antibiotics toward pathogens such as *Phytophthora cinnamomi*, e.g. mycorrhizae produced by *Leucopaxillus cerealis* var. *picena*. He also points out a well developed fungus sheath of ectomycorrhizae is less able to be penetrated by fungi such as *P. cinnamomi*. Note, antibiosis in laboratory media may be no indication of antibiosis in the rhizosphere (Bowen and Theodorou 1979).

Improved nutrition.   Improved nutrition caused by mycorrhizal development can affect disease indirectly, e.g. increased arginine content of mycorrhizal tobacco roots can suppress chlamydospore formation by *Thielaviopsis basicola* (Schönbeck and Schlösser 1976).

Compensation.   Roots lost by disease attack in a deficient soil will lead to reduced yield, but active mycorrhiza formation and hyphal growth into soil can compensate for such losses. Furthermore, this compensating tissue, i.e. the fungal hyphae, is *not* susceptible to the same diseases as the root and this provides a nice example of the many 'insurance' systems plants have developed during evolution. V. Bumbieris (personal communication) found mycorrhizal infection of *Pinus radiata* by *Rhizopogon luteolus* compensated for damage which would have occurred from *Phytophthora cryptogea* had the plant been non-mycorrhizal. (However, mycorrhizal infection in the absence of *P. cryptogea* gave an 80 per cent growth increase.)

Avoidance. Suberization or cutinization of newly formed roots confers a degree of disease protection but it also causes a very large decline in ion uptake by that part of the root. However, if the root has become mycorrhizal before suberization it can maintain an efficient uptake function via the mycorrhiza while enjoying the disease protection of suberization, e.g. in studies by Bowen, Bevege, and Mosse (1975) suberized uninfected laterals of the subtropical tree *Araucaria cunninghamii* exposed to ^{32}P orthophosphate had a count of 3.8 Ci/s, most of which was adsorption to the surface, i.e. there was no true phosphate uptake into the root. Similar mycorrhizae of the same size had 12.7 Ci/s, some three-quarters of which was true absorption.

It is also possible that mycorrhizal infection may elicit production of plant defence compounds against other micro-organisms for this occurs with orchid mycorrhizae (see Smith 1974) but this needs further study. Foster and Marks (1967) found enhanced polyphenol production in the root cortex of mycorrhizal *Pinus radiata* and Krupa and Fries (1971) found up to eight fold increases in volatile terpenes and sesquiterpenes (often fungistatic) with ectomycorrhizal infection of *Pinus sylvestris* by *Boletus variegatus*.

Bowen (1978) has reviewed interactions between VA mycorrhiza and nematodes. The studies of Schenck, Kinloch, and Dickson (1975) indicate a suppressive effect of some VA mycorrhizas on root-knot nematodes on soybeans.

Growth depression by mycorrhizae

The energy source for mycorrhizae is assimilates from the higher plant and although the mycorrhizal demand can be considerable, the benefits from the association usually far outweigh the energy cost by the fungus (see Bowen 1978). Detailed analyses of the energy relations of these symbioses have yet to be made; perhaps the plant may be able to compensate for an energy drain to the fungus by increasing photosynthesis. However there are cases where growth depression has been recorded with mycorrhizae:

   (i) temporary and relatively small depressions of plant growth (owing to fungal growth requirements) before the mycorrhiza response occurs have been recorded by a number of workers, e.g. on maize (Khan 1972).

  (ii) In situations where fertility is high enough to make mycorrhizae superfluous but not sufficiently high to inhibit mycorrhizal development seriously, depressions may occur due to the fungal demand for assimilate—data of Hepper (1977) shows some 17 per cent of the root weight can be fungal material in a heavily VA infected plant. Crush (1976) found decreases in growth of mycorrhizal *Medicago sativa* and *Trifolium hybridum* of 3.5–16.2 per cent.

 (iii) Mosse (1973b) found growth depressions of up to 40 per cent in mycorrhizal onion with supra-optimal additions of phosphate, caused by phosphate toxicity resulting from increased phosphate uptake by the mycorrhiza. This aspect of VA mycorrhiza may need careful attention in some soils because attempts to maximize production by adding considerable quantities of fertilizer could be counter-productive.

Mycorrhizae in mixed communities

A further dimension of plant and mycorrhizal ecophysiology is in maximizing productivity and stabilizing of mixed communities (as opposed to monocultures), e.g. natural communities such as tropical rain forests and savannah lands, or planted communities such as mixed pastures or the mixed cropping and inter-row cropping of tropical agricultural systems. The nutrient demands of two or more species growing concurrently are usually considerably in excess of that of a single species and a competitive situation often occurs. Any theoretical treatment or experimental study of competition for limited soil nutrients must necessarily include mycorrhizal systems.

In some communities competition is minimized by stratification of root zones of different species, e.g. ground-nut roots occupy the superficial layers of soil, while maize plants root at greater depth. In other cases different species have their major nutrient demand at different times and avoid intensive competition. However, competition intensifies when demands are concurrent and roots are occupying essentially the same location in soil.

In an excellent essay on plant interactions in mixed communities, Trenbath (1977) pointed out that nutrient competition between plant species begins when there is an overlap of depletion zones for nutrients by roots of different components of a mixed community. Therefore competition begins first for the more mobile ions, e.g. nitrate, sulphate, and potassium. For such ions a good correlation is observed between root abundance and nutrient absorbed. I deduce from this, in contrast to monoculture, that a situation where mycorrhizal fungus growth into soil is critical for uptake of mobile ions as well as poorly mobile ions—sparsely rooting species such as trees in grassland, or legumes in a grass pasture, and some grain legumes among maize, would be greatly disadvantaged in the absence of mycorrhizae. For this reason I believe mycorrhizae have been key factors in the evolution of mixed communities in nutrient-poor soils. However, the situation is not simple, for the densely rooting species are probably also mycorrhizal and it is possible that these could absorb luxury amounts of the elements in limited supply and disadvantage the sparsely rooting species even further. This situation would be alleviated if—as suggested by D.P. Janos (personal communication) and indicated by the data of Abbott and Robson (1977) on clover–mycorrhizal development in some densely rooting species decreases appreciably as their own nutritional requirements are met.

With non-mobile or poorly mobile ions depletion zones overlap only at high root densities and thus competition for these ions is probably delayed compared with that for more mobile ions. However such high root concentrations do occur with many grasses and herbs (Table 21.2) and again the mycorrhizal infection helps to place the sparsely rooting species at less of a disadvantage.

Few experimental studies have been made on the role of mycorrhizae in competition between plants. Crush (1974) found when *Trifolium repens* and *Lolium perenne* were grown together, mycorrhizae preferentially stimulated the growth of the legume in phosphorus deficient soil as predicted above. Hall (1978) found the growth of mycorrhizal white clover in ryegrass was forty times that of nonmycorrhizal clover in a phosphorus deficient soil. The ryegrass gave a poor response both to phosphorus and to VA mycorrhiza. Fitter (1977) found root competition and mycorrhiza infection gave a slight advantage to the grass *Holcus lanatus* but in combination these two factors depressed *Lolium perenne* considerably. Root length of *L. perenne* was reduced by 40 per cent by mycorrhiza infection in the mixture of grasses. Mycorrhizal inoculation of the combination of grasses gave considerable competitive advantage to *H. lanatus* in uptake of both potassium and phosphate.

Timing plays an important factor in competition for once a species gets a competitive advantage, differences are likely to be accentuated, e.g. better nutrition could lead to better growth and shading out of a competitor which then has less asimilate available for further root growth and mycorrhizal function. In the mycorrhizal context, a plant species which infects and responds quickly would thus have a competitive advantage. Bevege and Bowen (1975) observed differences of up to seven days between onion and clover in the time for infection with VA fungi. Few such studies have been performed but it is another important area where epidemiology and nutrition overlap.

Allelopathic effects, recognized between plant species, may also effect mycorrhizae and be important in competition and plant succession. One example of this is the reduction of infection in onion by swedes growing in the same pot (Hayman, Johnson, and Ruddlesdin 1975). Iqbal and Qureshi (1976) found the crucifer, *Brassica campestris* reduced VA infections of wheat in the field by 25 per cent and a similar loss in yield occurred. Morley and Mosse (1976) found that lupins and particularly lupin seed coats increased mycorrhizal infection on clover but many of the infections were morphologically abnormal—plant response was not studied. A substance from *Calluna vulgaris* can inhibit ectomycorrhizal fungi of *Picea abies* in heathland (Robinson 1972). D. P. Janos (personal communication) considered diminished mycorrhiza inoculum in sedge filled pastures in Costa Rica was responsible for failure of growth of tropical rain forest trees on the site but this may not be an allelopathic phenomenon. Obviously toxicities toward mycorrhizal fungi need further study, especially in mixed communities and with a range of mycorrhizal fungi.

A third facet of multiple cropping and mycorrhizal response is that of shading of one crop by another, e.g. coffee growing under rubber, cover crops and ground cover under trees, and shade tolerant legumes like *Calopogonium mucunoides*, cowpea, and mung bean intercropped with maize. Decreasing light intensity is well known to reduce mycorrhizal development and mycorrhizal response both with ectomycorrhizae (reviewed by Hacskaylo 1973) and VA mycorrhizae (Hayman 1974).

Conclusion

It is obvious that with mycorrhizal fungi, research workers have yet another tool and dimension to consider in their aims to increase or sustain productivity of forest and agricultural lands. The three-dimensional system of soil–mycorrhiza–plant properly managed, can transform our ideas of the production potential of nutritionally deficient soil and conserve dwindling and costly fertilizer reserves. The production crisis is nowhere more acute than in the tropics and it is possible that many of its nutritional problems can be alleviated by adding the particular growth and absorption characteristics of selected mycorrhizal fungi to that of the plant root. Variation exists between the mycorrhizal fungi in a number of characteristics and selection within them may be able to be effectively used in the many problem situations occurring in the tropics. I am not suggesting that mycorrhizal studies can replace the search for deficiency tolerant or alkaline tolerant plant varieties but that this search may be facilitated and augmented by appropriate mycorrhizal selection especially where plant varieties tolerant to deficiency also have undesirable characters, such as lodging in grain crops. Also, uptake by plant varieties selected in soils with high base status might conceivably be adapted to acid soils by inoculation with appropriate acidophilic mycorrhizal fungi. In the present state of our knowledge it seems the impact of mycorrhizal studies on breeding programmes will be on the uptake of nutrients and not on the use of nutrients once these are absorbed, but the effect of mycorrhizae on this last factor does need more study.

The value of mycorrhizal fungi is further enhanced when one considers their increased uptake of most of the ions needed to obtain efficient nitrogen fixation by legumes—the demand for nitrogen fixation is sometimes in excess of that required by the plant alone. This has extra importance in the tropics because of the grain legume programmes introduced to increase protein content of the diet. Furthermore, the legume is important in animal production—approximately half the world's permanent pastures and half the cattle population is in the tropics but productivity is low because beef is usually produced on soils of too low fertility for crops.

In this paper I have tried to emphasize ecological approaches and systems approaches to understanding mycorrhizal function in ion uptake, e.g. the role of mycorrhizae in competition and in the stable, tightly closed nutrient systems of natural savannahs and tropical rain forests. I have also suggested a wider vision of mycorrhizae in that they are an alternative or supplementary strategy for the plant in conditions deleterious to root growth, e.g. compensation for disease and counteracting toxicities. The tropics have relatively unique systems such as multiple cropping farming systems and in shifting cultivation. Much more research is needed on the role of particular mycorrhizal fungi in compatibility and competition in plant communities and their function in site recovery in shifting cultivation.

The mycorrhiza research worker in the tropics has an exciting task ahead of him. Not only can he make a significant contribution to producing more and better food but also he can make unique contributions to the science of understanding plant communities.

References

Abbott, L. K. and Robson, A. D. (1977). Growth stimulation of subterranean clover with vesicular arbuscular mycorrhizas. *Aust. J. agric. Res.* 28, 639–49.

Azcon, R., Barea, J. M., and Hayman, D. S. (1976). Utilization of rock phosphate in alkaline soils by plants inoculated with mycorrhizal fungi and phosphate-solubilizing bacteria. *Soil Biol. Biochem.* 6, 135–8.

Barea, J. M., Azcon, R., and Hayman, D. S. (1975). Possible synergistic interactions between *Endogone* and phosphate solubilizing bacteria in low phosphate soils. In *Endomycorrhizas* (eds. F. E. Sanders, B. Mosse, and P. B. Tinker) pp. 409–17. Academic Press, London.

Barley, K. P. (1970). The configuration of the root system in relation to nutrient uptake. *Adv. Agron.* 22, 159–201.

Barrow, N. J., Malajczuk, N., and Shaw, T. C. (1977). A direct test of the ability of vesicular arbuscular mycorrhiza to help plants take up fixed soil phosphate. *New Phytol.* 78, 269–76.

Baylis, G. T. S. (1975). The magnolioid root and mycotrophy in root systems derived from it. In *Endomycorrhizas* (eds. F. E. Sanders, B. Mosse, and P. B. Tinker) pp. 373–89. Academic Press, London.

Bevege, D. I. (1971). Vesicular-arbuscular mycorrhizas of *Araucaria*: aspects of their ecology and physiology and role in nitrogen fixation. Ph.D. Thesis, University of New England, Armidale, Australia.

—— and Bowen, G. D. (1975). *Endogone* strain and host plant differences in development of vesicular-arbuscular mycorrhizas. In *Endomycorrhizas* (eds. F. E. Sanders, B. Mosse, and P. B. Tinker) pp. 77–86. Academic Press, London.

—— —— and Skinner, M. F. (1975). Comparative carbohydrate physiology of ecto- and endomycorrhizas. In *Endomycorrhizas* (eds. F. E. Sanders, B. Mosse, and P. B. Tinker) pp. 149–74. Academic Press, London.

—— Richards, B. N., and Moore, A. W. (1978). Nitrogen fixation associated with conifers. Div. Soils, Div. Report 26, 1978. CSIRO Australia.

Bhat, K. K. S., Nye, P. H., and Baldwin, J. P. (1976). Diffusion of phosphate to plant roots in soil. IV. The concentration distance profile in the rhizosphere of roots with root hairs in a low-P soil. *Pl. Soil* 44, 63–72.

Bowen, G. D. (1968). Phosphate uptake by mycorrhizas and uninfected roots of *Pinus radiata* in relation to root distribution. *Proc. 9th Int. Cong. Soil Sci.* Vol. 2, pp. 219–28. Int. Soc. Soil Sci. and Angus and Robertson, Sydney.

—— (1973). Mineral nutrient relations of ectomycorrhizae. In *Ectomycorrhizae— their ecology and physiology* (eds. G. C. Marks and T. T. Kozlowski) pp. 151–205. Academic Press, New York.

—— (1978). Dysfunction and shortfalls in symbiotic response. In *Plant disease— an advanced treatise* (eds. J. G. Horsfall and E. B. Cowling) Vol. III, pp. 231–56. Academic Press, New York.

—— and Rovira, A. D. (1961). The effects of micro-organisms on plant growth. I. Development of roots and root hairs in sand and agar. *Pl. Soil* 15, 166–88.

—— —— (1976). Microbial colonization of roots. *A. Rev. Phytopathol.* 14, 121–44.

—— and Theodorou, C. (1979). Interactions between bacteria and ectomycorrhizal fungi. *Soil Biol. Biochem.* 10 119–26.

—— Bevege, D. I., and Mosse, B. (1975). Phosphate physiology of vesicular-arbuscular mycorrhizas. In *Endomycorrhizas* (eds. F. E. Sanders, B. Mosse, and P. B. Tinker) pp. 241–60. Academic Press, London.

—— Cartwright, B., and Mooney, J. R. (1974). Wheat root configuration under phosphate stress. In *Mechanisms of regulation of plant growth* (eds. R. L. Bieleski, A. R. Ferguson, and M. M. Creswell) pp. 121–5. Bull. 12, Royal Society New Zealand.

— Skinner, M. F., and Bevege, D. I. (1974). Zinc uptake by ecto- and endo-mycorrhizas and uninfected roots of conifers. *Soil Biol. Biochem.* 6, 141-4.

Cormack, R. G. H. (1962). Development of root hairs in angiosperms. II. *Bot. Rev.* 28, 446-64.

Crush, J. R. (1973). The effect of *Rhizophagus tenuis* mycorrhizas on ryegrass, cocksfoot and sweet vernal. *New Phytol.* 72, 965-73.

— (1974). Plant growth responses to vesicular-arbuscular mycorrhizas. VIII. Growth and nodulation of some herbage legumes. *New Phytol.* 73, 743-9.

— (1976). Endomycorrhizas and legume growth in some soils of the Mackenzie Basin, Canterbury, New Zealand. *N.Z. Jl agric. Res.* 19, 473-6.

Daft, M. J. and Okusanya, B. O. (1973a). Effect of *Endogone* mycorrhiza on plant growth V. Influence of infection on the multiplication of viruses in tomato, petunia and strawberry. *New Phytol.* 72, 975-83.

— — (1973b). Effect of *Endogone* mycorrhiza on plant growth. VI. Influence of infection on the anatomy and reproductive development in four hosts. *New Phytol.* 72, 1333-5.

El-Giahmi, A. A., Nicolson, T. H., and Daft, M. J. (1976). Endomycorrhizal fungi from Libyan soils. *Trans. Br. mycol. Soc.* 67, 164-9.

Fitter, A. H. (1977). Influence of mycorrhizal infection on competition for phosphorus and potassium by two grasses. *New Phytol.* 79, 119-25.

Foster, R. C. and Marks, G. C. (1967). Observations on the mycorrhizas of forest trees. II. The rhizosphere of *Pinus radiata* D. Don. *Aust. J. biol. Sci.* 20, 915-26.

Gadgil, R. L. and Gadgil, P. D. (1975). Suppression of litter decomposition by mycorrhizal roots of *Pinus radiata*. *N.Z. Jl For. Sci.* 5, 33-41.

Gerdemann, J. W. (1975). Vesicular-arbuscular mycorrhizae. In *The development and function of roots* (eds. J. G. Torrey and D. T. Clarkson) pp. 575-91. Academic Press, London.

Giles, K. L. and Whitehead, H. C. M. (1977). Reassociation of a modified mycorrhiza with the host plant roots (*Pinus radiata*) and the transfer of acetylene reduction activity. *Pl. Soil* 48, 143-52.

Gilmore, A. E. (1971). The influence of endotrophic mycorrhizae on the growth of peach seedlings. *J. Am. Soc. Hort. Sci.* 96, 35-8.

Gray, L. E. and Gerdemann, J. W. (1973). Uptake of sulphur-35 by vesicular arbuscular mycorrhizae. *Pl. Soil* 39, 687-9.

Hacskaylo, E. (1973). Carbohydrate physiology of ectomycorrhizae. In *Ectomycorrhizae—their ecology and physiology* (eds. G. C. Marks and T. T. Kozlowski) pp. 207-30. Academic Press, New York.

Haines, B. L. and Best, G. R. (1976). *Glomus mosseae*, endomycorrhizal with *Liquidambar styraciflua* L. seedlings retards NO_3, NO_2 and NH_4 nitrogen loss from a temperate forest soil. *Pl. Soil* 45, 257-61.

Hall, I. R. (1978). The effects of endomycorrhizas on the competitive ability of white clover. *N. Z. Jl agric. Res.* 21, 509-18.

Harley, J. L. (1969). *The biology of mycorrhiza*, 2nd edn. Leonard Hill, London.

Hattingh, M. J., Gray, L. E., and Gerdemann, J. W. (1973). Uptake and translocation of ^{32}P-labelled phosphate to onion roots by mycorrhizal fungi. *Soil Sci.* 116, 383-7.

Hayman, D. S. (1974). Plant growth responses to vesicular-arbuscular mycorrhiza. VI. Effect of light and temperature. *New Phytol.* 73, 71-80.

— and Mosse, B. (1972). Plant growth responses to vesicular-arbuscular mycorrhizas. III. Increased uptake of labile P from soil. *New Phytol.* 71, 41-7.

— Johnson, A. M., and Ruddlesdin, I. (1975). The influence of phosphate and crop species on *Endogone* spores and vesicular-arbuscular mycorrhizae under field conditions. *Pl. Soil* 43, 489-95.

Hepper, C. M. (1977). A colorimetric method for estimating vesicular-arbuscular mycorrhizal infection in roots. *Soil Biol. Biochem.* 9, 15–18.

Iqbal, S. H. and Qureshi, K. S. (1976). The influence of mixed sowing (cereals and crucifers) and crop rotation on the development of mycorrhiza and subsequent growth of crops under field conditions. *Biologia* 22, 287–98.

Jackson, N. E., Miller, R. H., and Franklin, R. E. (1973). The influence of vesicular-arbuscular mycorrhizas on uptake of ^{90}Sr from soil by soybeans. *Soil Biol. Biochem.* 5, 205–12.

Kerridge, P. C. (1978). Fertilization of acid tropical soils in relation to pasture legumes. In *Mineral nutrition of legumes in tropical and subtropical soils* (eds. C. S. Andrew and E. J. Kambrath) pp. 395–415, CSIRO, Melbourne, Australia.

Khan, A. G. (1972). The effect of vesicular-arbuscular mycorrhizal associations on the growth of cereals. I. Effects on maize growth. *New Phytol.* 71, 613–19.

—— (1974). The occurrence of mycorrhizas in halophytes, hydrophytes and xerophytes, and of *Endogone* spores in adjacent soils. *J. gen. Microbiol.* 81, 7–14.

Koucheki, H. K. and Read, D. J. (1976). Vesicular-arbuscular mycorrhiza in natural vegetation systems. II. The relationship between infection and growth in *Festuca ovina* L. *New Phytol.* 77, 655–66.

Krupa, S. and Fries, N. (1971). Studies on the ectomycorrhizae of pine. I. Production of volatile organic compounds. *Can. J. Bot.* 49, 1425–31.

Lamb, R. J. and Richards, B. N. (1971). Effect of mycorrhizal fungi on the growth and nutrient status of slash and radiata pine seedlings. *Aust. For.* 35, 1–7.

Marks, G. C. and Kozlowski, T. T. (1973). *Ectomycorrhizae—their ecology and physiology.* Academic Press, New York.

Marx, D. H. (1973). Mycorrhiza and feeder root diseases. In *Ectomycorrhizae—their ecology and physiology* (eds. G. C. Marks and T. T. Kozlowski) pp. 351–82. Academic Press, New York.

—— and Bryan, W. C. (1971). Influence of ectomycorrhizae on survival and growth of aseptic seedlings of loblolly pine at high temperature. *Forest Sci.* 17, 37–41.

—— —— and Davey, C. B. (1970). Influence of temperature on aseptic synthesis of ectomycorrhizae by *Thelephora terrestris* and *Pisolithus tinctorius* on Loblolly Pine. *Forest Sci.* 16, 424–31.

Mikola, P. (1970). Mycorrhizal inoculation in afforestation. *Int. Rev. For. Res.* 3, 123–96.

Moawad, M. (1979). Ecophysiology of v.a. mycorrhiza in the tropics. In *The soil-root interface* (eds. J. L. Harley and R. S. Russell) pp. 197–209. Academic Press, London.

Morley, C. D. and Mosse, B. (1976). Abnormal vesicular-arbuscular mycorrhizal infections in white clover induced by lupin. *Trans. Br. mycol. Soc.* 67, 510–13.

Morrison, T. M. (1962). Uptake of sulphur by excised beech mycorrhizas. *New Phytol.* 62, 44–9.

Mosse, B. (1972). The influence of soil type and *Endogone* strain on the growth of mycorrhizal plants in phosphate deficient soils. *Revue Écol. Biol. Sol* 9, 529–37.

—— (1973a). Advances in the study of vesicular-arbuscular mycorrhiza. *A. Rev. Phytopathol.* 11, 171–96.

—— (1973b). Plant growth response to vesicular-arbuscular mycorrhiza. IV. In soil given additional phosphate. *New Phytol.* 72, 127–36.

—— (1979). V.A. Mycorrhiza research in tropical soils. SOTA Report commissioned by US Agency for Internat. Devel. Publ. Univ. Hawaii Coll. Trop. Agric.

—— Hayman, D. S., and Arnold, D. J. (1973). Plant growth responses to vesicular-arbuscular mycorrhiza. V. Phosphate uptake by three plant species from P-deficient soils labelled with ^{32}P. *New Phytol.* 72, 809–15.

—— Powell, C. W., and Hayman, D. S. (1976). Plant growth responses to vesicular-arbuscular mycorrhiza. IX. Interactions between v.a. mycorrhiza, rock phosphate and symbiotic nitrogen fixation. *New Phytol.* 76, 331–42.

Nicolson, T. H. (1960). Mycorrhiza in the Gramineae. II. Development in different habitats, particularly sand dunes. *Trans. Br. mycol. Soc.* 43, 132–45.

Odum, H. T. (1970). *A tropical rain forest.* US Atomic Energy Commission, Washington.

Powell, C. Ll. (1975). Potassium uptake by endotrophic mycorrhizas. In *Endomycorrhizas* (eds. F. E. Sanders, B. Mosse, and P. B. Tinker) pp. 461–8. Academic Press, London.

Redhead, J. F. (1977). Endotrophic mycorrhizas in Nigeria: species of the Endogonaceae and their distribution. *Trans. Br. mycol. Soc.* 69, 275–80.

Reid, C. P. P. and Bowen, G. D. (1979). Effects of soil moisture on v.a. mycorrhiza formation and root development in *Medicago.* In *The soil-root interface* (eds. J. L. Harley and R. S. Russell) pp. 211–19. Academic Press, London.

Rhodes, L. H. and Gerdemann, J. W. (1975). Phosphate uptake zones of mycorrhizal and non-mycorrhizal onions. *New Phytol.* 75, 555–61.

Robinson, R. K. (1972). The production by roots of *Calluna vulgaris* of a factor inhibitory to growth of some mycorrhizal fungi. *J. Ecol.* 60, 219–24.

Rowell, D. L., Martin, M. W., and Nye, P. H. (1967). The measurement and mechanism of ion diffusion in soils. III. The effect of moisture content and soil solution concentration on the self-diffusion of ions in soils. *J. Soil Sci.* 18, 204–22.

Sanchez, P. A. (1976). *Properties and management of soils in the tropics.* John Wiley, New York.

Sanders, F. E. and Tinker, P. B. (1973). Phosphate flow into mycorrhizal roots. *Pestic. Sci.* 4, 385–95.

—— Mosse, B., and Tinker, P. B. (1975). *Endomycorrhizas.* Academic Press, London.

—— Tinker, P. B., Black, R. L. B., and Palmerley, S. M. (1977). The development of endomycorrhizal root systems I. Spread of infection and growth-promoting effects with four species of vesicular-arbuscular endophyte. *New Phytol.* 78, 257–68.

Sands, R. and Theodorou C. (1978). Water uptake by mycorrhizal roots of radiata pine seedlings. *Aust. J. Pl. Physiol.* 5, 301–9.

—— Greacen, E. L., and Gerard, C. J. (1979). Compaction of sandy soils in radiata pine forests. I. A penetrometer study. *Aust. J. Soil Res.* 17, 101–13.

Sanni, S. O. (1976). Vesicular-arbuscular mycorrhiza in some Nigerian soils: the effect of *Gigaspora gigantea* on the growth of rice. *New Phytol.* 77, 673–4.

Schenck, N. C., Kinloch, R. A., and Dickson, D. W. (1975). Interaction of endomycorrhizal fungi and root-knot nematodes on soybean. In *Endomycorrhizas* (eds. F. E. Sanders, B. Mosse, and P. B. Tinker) pp. 607–17. Academic Press, London.

Schönbeck, F. and Schlösser, E. (1976). Preformed substances as potential protectants. In *Physiological plant pathology* (eds. R. Heitefuss and P. H. Williams) pp. 653–78. Springer-Verlag, Berlin.

Skinner, M. F. and Bowen, G. D. (1974a). The uptake and translocation of phosphate by mycelial strands of pine mycorrhizas. *Soil Biol. Biochem.* 6, 53–6.

—— —— (1974b). The penetration of soil by mycelial strands of pine mycorrhizas. *Soil Biol. Biochem.* 6, 57–61.

Smith, S. E. (1974). Mycorrhizal fungi. *Crit. Rev. Microbiol.* 2, 275-313.

Sutton, J. C. (1973). Development of vesicular-arbuscular mycorrhizae in crop plants. *Can. J. Bot.* 51, 2487-93.

— and Sheppard, B. R. (1976). Aggregation of sand dune soil by endomycorrhizal fungi. *Can. J. Bot.* 54, 326-33.

Theodorou, C. (1968). Inositol phosphates in needles of *Pinus radiata* and the phytase activity of mycorrhizal fungi. *Proc. 9th Int. Cong. Soil Sci.*, Vol. 3, pp. 483-93. Int. Soc. Soil Sci. and Angus and Robertson, Sydney.

— (1978). Soil moisture and the mycorrhizal association of *Pinus radiata* D. Don. *Soil Biol. Biochem.* 10, 33-7.

— and Bowen, G. D. (1971). Influence of temperature on the mycorrhizal associations of *Pinus radiata. Aust. J. Bot.* 19, 13-20.

Trappe, J. M. (1977). Selection of fungi for ectomycorrhizal inoculation in nurseries. *A. Rev. Phytopathol.* 15, 203-22.

Trenbath, B. R. (1977). Plant interactions in mixed crop communities. In *Multiple cropping* (ed. M. Stelley) pp. 129-69. American Soc. Agron. Special Publication 27. Madison, Wis.

Vozzo, J. A. and Hacskaylo, E. (1971). Inoculation of *Pinus caribaea* with ectomycorrhizal fungi in Puerto Rico. *Forest Sci.* 17, 239-45.

Went, F. W. and Stark, N. (1968). The biological and mechanical role of soil fungi. *Proc. natn. Acad. Sci. U.S.A.* 60, 497-504.

Williamson, B. and Alexander, I. J. (1975). Acid phosphatase localized in the sheath of beech mycorrhiza. *Soil Biol. Biochem.* 7, 195-8.

22 The role of mycorrhizal symbiosis in the nutrition of tropical plants

R. Black

This paper is concerned with the distribution and significance of mycorrhizae in tropical plants, taking as the starting point fundamental mycorrhiza research concerned primarily with temperate host species and reviewed by Harley (1969), Nicolson (1967), Gerdemann (1968), Mosse (1973), Marks and Kozlowski (1973), and Tinker (1975). In Table 22.1 the known types of mycorrhizae are listed according to the scheme of Lewis (1973).

Table 22.1. Clasification of mycorrhizae (after Lewis 1973)

Mycorrhiza	Hosts	Mode of nutrition of host
Sheathing	Seed plants: trees and shrubs	Photoautotrophic
Vesicular-arbuscular	Seed plants: herbaceous and woody Pteridophytes (sporophyte) Bryophytes: liverworts	Photoautotrophic
Ericaceous: Ericoid Arbutoid	Ericales	Photoautotrophic
Orchidaceous	Orchidaceae Liverworts	Saprotrophic or nectrotrophic, at least in early growth
Mycorrhizae of other saprophytic plants*	Pyrolaceae, Gentianaceae, Liliaceae, Triuridaceae, Polygalaceae, Burmanniaceae	Entirely saprotrophic or necrotrophic
Miscellaneous	e.g. Cistaceae	

*Endophytes uncertain.

The hosts of sheathing mycorrhizae ('ectomycorrhizae' and related forms) originate primarily in temperate and Mediterranean regions of the northern hemisphere, but are widely planted in the tropics. Vesicular-arbuscular mycorrhizae are cosmopolitan and almost universal in host range. The orchid flora of the tropics is especially rich; most entirely non-green orchids are tropical. A large proportion of saprophytic Angiosperms of other families are indigenous to the tropics; the tropical families Burmanniaceae and Triuridaceae depend completely on the mycorrhizal fungus for their carbon nutrition. The Ericaceae are of little significance in the tropics and will not be considered further.

Mycorrhizae of tropical orchids and other saprophytic plants

This section deals first with green orchids, then entirely achlorophyllous Angiosperms, both groups of plants receiving organic carbon compounds from a saprobic fungal endophyte.

The physiology and ecology of orchid mycorrhizae has been reviewed by Burgeff (1959), Furman and Trappe (1971), Harley (1969), Sanford (1974), Withner (1974), and Warcup (1975). Green and saprophytic orchids are unique in that penetration of the mycorrhizal fungus into the swollen seed is necessary for continued germination. Commercial interest in optimizing germination of the tiny orchid seed, either symbiotically or aseptically, has given impetus to the investigation of the precise role of the fungus in the nutrition of the host (Warcup 1975). In temperate species there may be an extended phase of seedling development from the protocorm to the first leaf where the plant is clearly heterotrophic. In tropical species the protocorm may green rapidly but Hadley (1969) established that mycorrhizal infection could enhance the growth of green protocorms. Significantly, cellulose as a carbon source greatly enhanced protocorm growth; whereas if glucose was used its supply had to be limited to prevent parasitism of the orchid by the fungus.

After germination there is wide variation in the dependence of the orchid on the mycorrhiza, related to the development of photosynthetic capacity (Furman and Trappe 1971). Nutrient exchange in the orchid mycorrhizal symbiosis involves transfer of carbohydrate from the fungus to the orchid, and the fungus, being heterotrophic for nitrogen, receives amino acids manufactured by the host. Some orchid endophytes require vitamins as well (Sanford 1974; Powell and Arditti 1975).

The orchid should be regarded as parasitizing the fungus necrotrophically since digestion of the fungus is a regular feature of the interaction (Lewis 1973). However, the digestive process is not necessarily essential to nutrient exchange but should be regarded as a method of preventing parasitism by the fungus (Hadley and Williamson 1971), along with antifungal phytoalexins (Withner 1974).

The contribution made by mycorrhizae to the carbon nutrition of a green orchid which retains the endophyte is generally unclear. One should note that achlorophyllous individuals can develop in species generally green, which suggests a flexible mechanism for using fungal saprobism to make up for conditions unfavourable for photosynthesis.

The tropics are characterized by the epiphytic habit among Angiosperms, restricted to Pteridophytes, Bryophytes, and lichens in temperate regions. Epiphytic orchids are very numerous in the tropics. Bark and moss may be rich in soluble organic compounds and Sanford (1974) has commented upon the necessity for orchid establishment in this habitat.

It has been claimed that all epiphytic orchids may damage the supporting tree through the parasitism of the fungal endophyte (see Harley 1969). This suggestion has been criticized recently (Sanford 1974) but the background observations do demonstrate the existence and potential importance of a prolonged mycelial phase in this habitat.

The extension of saprophytism to the mature orchid is accompanied by reduction of roots as well as in green leaf area. In extreme cases roots may be lacking in a completely achlorophyllous species with the endophyte in the rhizome (Furman and Trappe 1971). Typically, seed germination of terrestrial orchids is subterranean. Many achlorophyllous orchids grow entirely underground for several years before the inflorescence emerges.

Tropical species of saprophytes are also found in the families Pyrolaceae, Gentianaceae, Burmanniaceae, Polygalaceae, and Triuridaceae. The Pyrolaceae is the only family where most of the species are not tropical.

The chlorophyll-lacking plant is not really saprophytic but parasitic on its mycorrhizal fungus (Furman and Trappe 1971). It has been shown that some orchids and members of the Pyrolaceae are connected via their endophytes to the roots of photosynthetic plants. In the case of the Pyrolaceae it has been demonstrated conclusively that there is transfer of nutrients from the green plant via the fungus to the so-called saprophyte. It is significant that those achlorophyllous orchids presumed to be epiparasitic have the facultative pathogen *Armillaria* as endophyte, rather than *Rhizoctonia*.

Richards (1952) has drawn attention to the peculiar distribution pattern of tropical achlorophyllous plants. They occur in the darkest regions of the forest in permanently wet conditions, but such conditions for excluding competition from photosynthetic plants cannot explain their very scattered distribution. Furman and Trappe (1971) suggest that epiparasitism may be widespread and the distribution of host species may determine their location.

Sheathing mycorrhizae

Included in this category are ectomycorrhizae, ectendomycorrhizae, and pseudomycorrhizae. Compared to the range of ectomycorrhizal seed plants of temperate and Mediterranean regions the number of tropical plants known with this type of mycorrhiza is much more limited (Table 22.2).

Table 22.2. Families of seed plants having tropical species with sheathing mycorrhizae

Euphorbiaceae	cosmopolitan
Pinaceae	
Fagaceae	principally temperate and Mediterranean
Myrtaceae	*Eucalyptus*: a few tropical species *Camponanesia*
Dipterocarpaceae	S.E. Asia
Caesalpiniaceae	
Sapindaceae	Pan-tropical

From Meyer (1973) and Redhead (Chapter 16).

The Pinaceae include some tropical *Pinus* species of montane forests, *P. caribaea, P. merkusii, P. kesiya*, with a number of species indigenous to the drier subtropics. There are a few tropical species of the Fagaceae, Euphorbiaceae, and Myrtaceae which form sheathing mycorrhizae. Reports of sheathing mycorrhizae in tropical families have so far been restricted to the Caesalpiniaceae, Dipterocarpaceae, and Sapindaceae (Meyer 1973).

An important feature of tropical vegetation at the present time, however, are the plantations of exotic Pinaceae and *Eucalyptus*. Mycorrhizal inoculation is a key part in this afforestation.

Meyer (1974) has provided an illuminating review of the physiology of sheathing mycorrhizae, particularly from the standpoint of the fungus. The sheathing mycorrhiza should be regarded as a dual parasitism, with the fungus an ecologically obligate symbiotic biotroph in the terminology of Lewis (1973). The other side of the parasitism/symbiosis is increased uptake of nutrients by the host, the mechanisms of which are reviewed by Bowen (1973) and Cooke (1977).

Where there is a true ectomycorrhiza probably the host is always mycotrophic. As a result literature on sheathing mycorrhizae in the tropics has concentrated on the availability and application of inoculation, which is inappropriate to this review. However, there are various implications of the fact that the sheathing mycorrhiza is essentially a feature of the temperate forest that deserve consideration.

Firstly, sheathing mycorrhizae exploit the leaf litter layers, obtaining nutrients released by the activity of saprobic micro-organisms. The realization of the climatic potential for high productivity in a tropical rain forest is dependent upon rapid cycling of nutrients, most of which are locked up in the biomass of the trees. The work of Gadgil and Gadgil (1971, 1975) suggests that sheathing mycorrhizae actually suppress litter decomposition, possibly by removing nutrients required by decomposing micro-organisms. This factor may be particularly important in the tropics in view of the contribution of litter decomposition to nutrient availability. Table 22.3 compares the amount of standing litter with the annual leaf fall (proportional to biomass) in different forest types.

Table 22.3. Litter and leaf fall in different forest types

	Standing litter (t/ha)	Annual leaf fall (t/ha)	Approx. ratio litter: leaf fall
Taiga (USSR)	30–45	2	15–20
High oak forest (USSR)	15	3.8	4
Upper montane rain forest (Jamaica)	11	5	2.2
Lowland rain forest Asia	2	20*	0.1
Trinidad	4.2	6.8	0.6

*This high figure may result from a seasonal leaf fall having been given undue weighting on an annual basis.
From Satchell (1974) and Tanner (1977).

Attention has been drawn to the need for more information about nitrate and ammonium uptake by mycorrhizae (Bowen 1973). Bowen suggested that there should be selection for nitrate requiring fungi, in view of the presence of nitrate from nitrification in neutral and alkaline soils, and since nitrate fertilizers are being used in forestry. However, mineralization can be very rapid in the tropics, with nitrification not keeping up, even in neutral soils, so that an ammonium requiring mycorrhizal fungus may be an advantage in certain conditions.

Finally, there is the question of linking sheathing mycorrhizae and nitrogen fixation. Among temperate plants *Alnus* (Betulaceae) is probably the only nitrogen fixing ectomycorrhizal host (non-Rhizobial nodules). However, several genera of the Caesalpiniaceae in the tropics are known to be ectomycorrhizal, and the list is likely to be extended. The significance of this dual symbiosis in nutrient cycles should receive more attention as it may be important in savannah as well as in forest. Also of interest is the work of Giles and Whitehead (1976) who produced mycelium of *Rhizopogon* containing nitrogen fixing cells of *Azotobacter* in the hyphae, which has now been used to infect *Pinus* seedlings (Giles 1977).

Vesicular-arbuscular (VA) mycorrhizae

Forest dominates the natural vegetation of the tropics with VA mycorrhizae in most plant forms—Angiosperms, Gymnosperms (Cycadales, Podocarpaceae), Pteridophytes, liverworts. VA mycorrhizae also occur in agricultural plants and other species characteristic of human civilization. Finally, there are many more recent introductions with VA mycorrhizae including exotic conifers (Cupressaceae, Araucariaceae).

Most of the truly tropical families which lack VA mycorrhizae in all or most of their members are xerophytic, e.g. Capparaceae, Nyctaginaceae, Portulacaceae, Phytolaccaceae (Gerdemann 1968; Khan 1974). VA mycorrhizae are also lacking in a number of cosmopolitan hydrophytic and halophytic families.

Although VA mycorrhizae are widespread, not all plant hosts are necessarily mycotrophic. Mycotrophy is largely a feature of woody and herbaceous plants lacking root hairs, whereas a response to VA mycorrhizal infection in, for example, grasses, can only be induced when available phosphorus is vanishingly low (Baylis 1972).

The explanation of the effects of VA mycorrhizae are summarized by Tinker (1975), and principally involves phosphorus absorption. Adverse, as well as beneficial effects of VA mycorrhizae on the host should not be neglected (Crush 1975; Black and Tinker 1977). Like the fungi of sheathing mycorrhizae the endophyte is an obligately symbiotic biotroph.

Tropical plants can now be considered specifically. In surveys of lowland rain forest (LRF), with indigenous and exotic trees, invariably a high proportion of species are found to be VA mycorrhizal—Nigeria (Redhead 1968), Java (Richards 1952), Costa Rica (Janos 1975). Similarly, a great number of species of the Cerrado vegetation of Brazil host VA mycorrhizae (Thomazini 1974). Janos had noticed that most of the species in his study lacked root hairs, suggesting mycotrophy. The only non-mycotrophic group of plants were primary colonizers of burnt ground, flood sites, etc. One of the most remarkable types of LRF in the Caribbean region is that developed on white sands, and here VA mycorrhizae are very important.

A major role of VA mycorrhizae is in seedling survival and many species may be obligately mycorrhizal in the seedling stage. Large seed reserves do not obviate the need for mycorrhizal infection—in fact they are necessary to achieve an adequate root system rapidly in order to increase the chances of infection (Janos 1975).

Individual tropical tree crops have received attention: Cacao (Laycock 1945), Litchi (Pandey and Misra 1971, 1975), Avocado (Hall and Finch 1974), *Bactris* (Palmae) (Janos 1976). Janos noted the need for VA mycorrhizae in the establishment of the palm. Other workers have concentrated on the effects of mycorrhizae on established plants or merely noted their presence.

There has been little experimental work on mycorrhizae of tropical annual crops, with the exception of legumes which will be considered later. VA mycorrhizae do not seem essential to the establishment of such crops but inoculation experiments have shown that the symbiosis may increase yield (Johnston 1949; Khan 1972; Ramirez, Mitchell, and Schenk 1975; Sanni 1976a,b). A neglected area of experimental work, however, has been the relationship between the development of infection and the development of the root system and of nutrient demand. Also we do not know the significance of normal infection levels in a crop resulting from indigenous inoculum rather than from inoculation.

It is almost a cliché in the mycorrhizal world now that the potential for the exploitation of VA mycorrhizae in agriculture is much greater in the tropics than in temperate countries. This statement recognizes the generally low availability of phosphorus in tropical soils which can be attributed to the following factors:

(i) Intrinsically low available phosphorus in agricultural soils derived by forest clearance. Sanchez and Buol (1975) estimated that leached, weathered soils in the tropics occupy 51 per cent of the land surface.

(ii) Many of the Red Latosols are phosphate-fixing, e.g. Red Bauxitic soils of the Caribbean. Any fertilizer phosphorus surplus to crop requirements is fixed rapidly and does not augment available phosphorus.

(iii) In general there has not been a long history of phosphorus fertilizer use that could have raised available phosphorus where possible.

(iv) Many Third-World countries lack facilities for manufacturing soluble phosphate fertilizers, and importation is becoming a prohibitively expensive drain on foreign exchange.

Such factors have stimulated interest in controlled-release fertilizers (e.g. Novoa and Nunez (1974) in the Caribbean) and the use of insoluble rock phosphate or its derivatives. The solubility of rock phosphate is too low to give significant yield response without mycorrhizae (Mosse, Powell, and Hayman 1976; Mosse 1977) but many agronomists still may not be aware of VA mycorrhizae. There are other cases of anomalous results which could possibly be explained by VA mycorrhizae. For instance, in Guyana it has been noted that plants may do well on soils with very little available phosphorus, provided they are well structured to permit extensive root exploitation (Hardy 1974).

Finally, one should note the importance of tropical legumes as doubly symbiotic plants. The following points should be considered:

(i) Legumes are very important sources of protein in the tropics; they are usually VA mycorrhizal (Possingham, Obbink, and Jones 1971).

(ii) Phosphorus stimulates nodulation. Together with water, phosphorus is the major factor limiting performance of tropical legumes.

(iii) Tropical legumes are characteristically sparse in root hairs and so potentially mycotrophic (Crush 1974), or known to be so (Daft and El-Giahmi 1976).

(iv) VA mycorrhizae in legumes are tolerant of high phosphorus applications (Abbott and Robson 1977).

(v) Tropical mycorrhizal legumes respond to rock phosphate (Azcon, Barea, and Hayman 1976; Mosse 1977; Mosse *et al.* 1976).

Conclusions

In tropical plant communities orchidaceous, sheathing, and VA mycorrhizae all play important roles in the nutrition of their hosts. One should contrast the function of sheathing and VA mycorrhizae with that of the orchidaceous type. It is worth noting, however, various contrasting features of the 'phosphorus mycorrhizae' which relate to adaptation to rather different conditions (Table 22.4).

Table 22.4. Some contrasting features of sheathing and VA mycorrhizae

	Sheathing	Vesicular-arbuscular
Mode of nutrition of fungus	Ecologically obligate symbiotic biotroph	Obligate symbiotic biotroph
Main effect on host nutrition	Increased P uptake (increased N uptake)	Increased P uptake
Distribution of hosts	Mainly outside tropics	Worldwide. Main mycorrhiza in tropics
Distribution of inoculum	Exotic with respect to tropics	Indigenous
Source of phosphorus in tropics	Litter	Soil, fertilizers, litter
Exploitation of soil	Low root density, sheath, mycelial strands	High root density, hyphae
Phosphorus reservoir	Sheath	Only limited P reservoir, if any
Nitrogen fixation	Few (?) leguminous trees	Numerous herbage and pulse legumes. (Rhizosphere, phyllo-plane and leaf nodule N-fixation)

Acknowledgements

I wish to thank my colleagues in the Botany Department, UWI, Mona for helpful discussion during the writing of this paper. My participation in the Workshop at which this paper was presented was made possible through the generous financial support of the Commonwealth Fund for Technical Co-operation. I also thank Mrs J. Wynter and Mrs E. Budhai for typing the manuscript.

References

Abbott, L. K. and Robson, A. D. (1977). Growth stimulation of subterranean clover with vesicular-arbuscular mycorrhizas. *Aust. J. agric. Res.* **28**, 639–49.

Azcon, R., Barea, J. M., and Hayman, D. S. (1976). Utilisation of rock phosphate in alkaline soils by plants inoculated with mycorrhizal fungi and phosphate-solubilizing bacteria. *Soil Biol. Biochem.* **8**, 135–8.

Baylis, G. T. S. (1972). Fungi, phosphorus and the evolution of root systems. *Search* **3**, 257–8.

Black, R. L. B. and Tinker, P. B. (1977). Interaction between effects of vesicular-arbuscular mycorrhiza and fertilizer phosphorus on yields of potatoes in the field. *Nature, Lond.* 267, 510-11.

Bowen, G. D. (1973). Mineral nutrition of ectomycorrhizae. In *Ectomycorrhizae: their ecology and physiology* (eds. G. C. Marks and T. T. Kozlowski) pp. 151-205. Academic Press, New York.

Burgeff, H. (1969). Mycorrhiza of orchids. In *The orchids: a scientific survey* (ed. C. L. Withner). Ronald Press, New York.

Cooke, R. (1977). *The biology of symbiotic fungi.* Wiley, London.

Crush, J. R. (1974). Plant growth responses to vesicular-arbuscular mycorrhiza. VII. Growth and nodulation of some herbage legumes. *New Phytol.* 73, 743-9.

—— (1975). Occurrence of endomycorrhizas in soils of the Mackenzie Basin, Canterbury, New Zealand. *N.Z. Jl agric. Res.* 18, 361-4.

Daft, M. J. and El-Giahmi, A. A. (1976). Studies on nodulated and mycorrhizal peanuts. *Ann. appl. Biol.* 82, 273-6.

Furman, T. E. and Trappe, J. M. (1971). Phylogeny and ecology of mycotrophic achlorophyllous Angiosperms. *Q. Rev. Biol.* 46, 219-25.

Gadgil, R. L. and Gadgil, P. D. (1971). Mycorrhiza and litter decomposition. *Nature, Lond.* 233, 133.

—— —— (1975). Suppression of litter decomposition by mycorrhizal roots of *Pinus radiata. N.Z. Jl For. Sci.* 5, 33-41.

Gerdemann, J. W. (1968). Vesicular-arbuscular mycorrhiza and plant growth. *A. Rev. Phytopathol.* 6, 397-418.

Giles, K. L. (1977). Transfer of nitrogen-fixing ability to eukaryote cells. In *Plant cell and tissue culture, Symposium,* Ohio State University, 6-9 September 1977.

—— and Whitehead, H. (1976). Uptake and continued metabolic activity of *Azotobacter* within fungal protoplasts. *Science, N.Y.* 193, 1125-6.

Hadley, G. (1969). Cellulose as a carbon source for orchid mycorrhiza. *New Phytol.* 68, 933-9.

—— and Williamson, B. (1971). Analysis of post-infection growth stimulus in orchid mycorrhiza. *New Phytol.* 70, 445-55.

Hall, J. B. and Finch, H. C. (1974). Mycorrhiza in roots of Avocado: effect upon chemotaxis of *Phytophthora cinnamomi* zoospores. *Proc. Am. Phytopathol. Soc.* 1, 86.

Hardy, F. (1974). Root room. *Trop. Agric., Trin.* 51, 272-8.

Harley, J. L. (1969). *The biology of mycorrhiza,* 2nd edn. Leonard Hill, London.

Janos, D. P. (1975). Effects of vesicular-arbuscular mycorrhizae on lowland tropical rainforest trees. In *Endomycorrhizas* (eds. F. E. Sanders, B. Mosse, and P. B. Tinker). Academic Press, London.

—— (1976). Vesicular-arbuscular mycorrhizae affect the growth of *Bactris gasipes. Principes* 21, 12-18.

Johnston, A. (1949). Vesicular-arbuscular mycorrhizas in Sea Island cotton and other tropical plants. *Trop. Agric., Trin.* 26, 118-21.

Khan, A. G. (1972). The effect of vesicular-arbuscular mycorrhizal associations on growth of cereals. I. Effect on maize growth. *New Phytol.* 71, 613-19.

—— (1974). The occurrence of mycorrhizas in halophytes, hydrophytes and xerophytes, and of *Endogone* spores in adjacent soils. *J. gen. Microbiol.* 81, 7-14.

Laycock, D. H. (1945). Preliminary investigations into the function of the endotrophic mycorrhiza of *Theobroma cacao* L. *Trop. Agric., Trin.* 22, 77-80.

Lewis, D. H. (1973). Concepts in fungal nutrition and the origin of biotrophy. *Biol. Rev.* 48, 261-78.

Marks, G. C. and Kozlowski, T. T. (1973). *Ectomycorrhizae: their ecology and physiology*. Academic Press, New York.

Meyer, F. H. (1973). Distribution of ectomycorrhizae in native and man-made forests. In *Ectomycorrhizae: their ecology and physiology* (eds. G. C. Marks and T. T. Kozlowski) pp. 79-105. Academic Press, New York.

—— (1974). Physiology of mycorrhiza. *A. Rev. Pl. Physiol.* 25, 567-86.

Mosse, B. (1973). Advances in the study of vesicular-arbuscular mycorrhiza. *A. Rev. Phytopathol.* 11, 171-96.

—— (1977). Plant growth responses to vesicular-arbuscular mycorrhiza. X. Responses of *Stylosanthes* and maize to inoculation in unsterile soils. *New Phytol.* 78, 277-88.

—— Powell, C. L., and Hayman, D. S. (1976). Plant growth responses to vesicular-arbuscular mycorrhiza. IX. Interactions between VA mycorrhiza, rock phosphate and symbiotic nitrogen fixation. *New Phytol.* 76, 331-42.

Nicolson, T. H. (1967). Vesicular-arbuscular mycorrhiza—a universal plant symbiosis. *Sci. Prog., Lond.* 55, 561-81.

Novoa, H. F. C. and Nunez, R. (1974). Efficiency of five phosphate fertilizer sources in soils with different phosphate fixing capacities. *Trop. Agric., Trin.* 5, 235-45.

Pandey, S. P. and Misra, A. P. (1971). *Rhizophagus* in mycorrhizal association with *Litchi sinensis* Sonn. *Mycopath. Mycol. appl.* 45, 337-54.

—— —— (1975). Mycorrhizas in relation to growth and fruiting of *Litchi sinensis* Sonn. *J. Indian Bot. Soc.* 54, 280-93.

Possingham, J. V., Obbink, J. G., and Jones, R. K. (1971). Tropical legumes and vesicular-arbuscular mycorrhiza. *J. Aust. Inst. agric. Sci.* 37, 160-1.

Powell, K. B. and Arditti, J. (1975). Growth requirements of *Rhizoctonia repens* M32. *Mycopathologia* 53, 161-7.

Ramirez, B. N., Mitchell, D. J., and Schenk, N. C. (1975). Establishment and growth effects of three vesicular-arbuscular mycorrhizal fungi on papaya. *Mycologia* 67, 1039-41.

Redhead, J. F. (1968). Mycorrhizal associations in some Nigerian forest trees. *Trans. Br. Mycol. Soc.* 51, 377-87.

Richards, P. W. (1952). *The tropical rain forest. An ecological study*. Cambridge University Press.

Sanchez, P. A. and Buol, S. W. (1975). Soils of the tropics and the world food crisis. In *Food: politics, economics, nutrition and research* (ed. P. H. Abelson). American Association for the Advancement of Science, Washington, DC.

Sanford, W. W. (1974). The ecology of orchids. In *The orchids: scientific studies* (ed. C. L. Withner). Wiley, New York.

Sanni, S. O. (1976a). Vesicular-arbuscular mycorrhiza in some Nigerian soils and their effects on the growth of cowpea (*Vigna unguiculata*), tomato (*Lycopersicon esculentum*) and maize (*Zea mays*). *New Phytol.* 77, 667-73.

—— (1976b). Vesicular-arbuscular mycorrhiza in some Nigerian soils: the effects of *Gigaspora gigantea* on the growth of rice. *New Phytol.* 77, 673-4.

Satchell, J. E. (1974). Litter-interface of animate/inanimate matter. In *Biology of plant litter decomposition* (eds. C. H. Dickinson and G. J. F. Pugh). Academic Press, London.

Tanner, E. V. J. (1977). Mineral cycling in Montane Rain Forests in Jamaica. Ph.D. Thesis, University of Cambridge.

Thomazini, L. I. (1974). Mycorrhiza in plants of the 'Cerrado'. *Pl. Soil* 41, 707-11.

Tinker, P. B. (1975). Effects of vesicular-arbuscular mycorrhizas on higher plants. *Symposium, Society for Experimental Biology*, Vol. 29, pp. 325-49.

Warcup, J. H. (1975). Factors affecting symbiotic germination of orchid seed. In *Endomycorrhizas* (eds. F. E. Sanders, B. Mosse, and P. R. Tinker). Academic Press, London.

Withner, C. L. (1974). Developments in orchid physiology. In *The orchids: scientific studies* (ed. C. L. Withner). Wiley, New York.

23 Ecophysiology of vesicular-arbuscular mycorrhiza

M. Moawad

Abstract

Phosphorus in tropical soils, predominantly acidic latosols, is largely fixed as ferri- and/or aluminium phosphate and is not available to plants. Improved growth of crops in the presence of VA mycorrhiza has been demonstrated many times and is greatest in soils of low fertility containing little available phosphate. The aim of our investigations was to find out which agronomic methods should be employed to exploit fully VA mycorrhiza in phosphorus nutrition of tropical crops. We studied several factors controlling the activity of the fungi, especially soil temperature, soil pH, water regime, and the interactions of fertilizer application and light intensity with soil temperature.

Effect of soil temperature

Nyabyenda (1977) found that with insoluble phosphate, *Eupatorium odoratum* without mycorrhiza was unaffected by soil temperature between 20 and 35 °C and phosphorus uptake was uniformly low. Mycorrhizal plants showed a very strong response with a steep rise from 20 to 30 °C and a fall from 30 to 35 °C. Growth of mycorrhizal plants was closely correlated with phosphorus uptake which was much higher than with non-mycorrhizal plants. When soluble phosphate was supplied, mycorrhiza showed a smaller effect on growth and phosphorus uptake. At the lowest temperature (20 °C) the non-mycorrhizal plants grew better and took up more phosphate than mycorrhizal, whereas at the highest temperatures (30 and 35 °C), the mycorrhizal plants were always superior, although the effect was not as striking as with insoluble phosphate. Somewhat similar results have been obtained by other authors with different plants.

In contrast to *E. odoratum*, Nyabyenda (1977) found that non-mycorrhizal plants of *Guizotia abyssinica* and *Sorghum bicolor* did respond to temperature in the presence of insoluble $Ca_5(PO_4)_3OH$, *G. abyssinica* less than *S. bicolor*. The optimum soil temperature for both plants with and without mycorrhiza was 30 °C. At temperatures of 25 °C and higher, both species absorbed considerable amounts of phosphorus from insoluble compounds in the absence of mycorrhiza, sorghum more efficiently than *Guizotia*, whilst mycorrhiza further increased phosphorus uptake and growth, particularly at 25 and

30 °C. There is an indication that two mechanisms with different response to temperature are involved in phosphorus uptake since the increase in phosphorus uptake and growth from 20 to 25 °C was much steeper in mycorrhizal than in non-mycorrhizal plants. The much sharper drop in phosphorus uptake and growth of sorghum from 30 to 35 °C also points to differences between non-mycorrhizal and mycorrhizal plants. The adaption of *G. abyssinica* to relatively low temperatures led to very poor growth at 35 °C. Temperature seems to affect the physiological activity of the mycorrhiza more than its development, since plants kept at the optimum temperature (30 °C) for 20 days before being moved to lower or higher temperatures showed the same influence of temperature on growth and phosphorus uptake as plants kept from the beginning at the different temperatures.

In an experiment with diurnally changing temperatures (30/20, 35/30, 35/25, 40/25) 35 °C proved not to be harmful to the development and physiological activity of the mycorrhiza if this was given only during the day, followed by night temperatures of 25 or 30 °C.

Effect of light intensity

Some authors have reported contradictory results as to the effect of light on mycorrhiza activity. Nyabyenda (1977) found that shading had no effect on non-mycorrhizal, but decreased the efficiency of mycorrhizal *E. odoratum* plants. At 30 °C, shading had less effect on mycorrhizal plants (growth and phosphorus uptake) than at 25 and 35 °C.

Effect of soil pH

Graw (1978) found in two species of the same family a different influence of soil pH on the effect of VA mycorrhiza on growth and phosphorus uptake when phosphorus was supplied as $Ca_5(PO_4)_3OH$.

Mycorrhizal *Tagetes minuta* grew better and took up more phosphate at all pHs than non-mycorrhizal, the difference being greatest at pH 4.3. The availability of $Ca_5(PO_4)_3OH$ decreases with rising soil pH; correspondingly, the growth and phosphorus uptake of mycorrhizal *T. minuta* decreased at higher pH values.

In contrast to *T. minuta*, mycorrhizal *G. abyssinica* grew better and took up more phosphate at pH 5.6 and 6.6 than at pH 4.3. Non-mycorrhizal *G. abyssinica* grew best at pH 4.3; at this pH mycorrhizal plants were much inferior to non-mycorrhizal. The reason for this is not clear. Increasing the soil pH to 6.6 decreased the dry weight and phosphorus content of non-mycorrhizal plants, but the positive effect of the mycorrhiza was slightly increased.

The different reactions to soil pH in presence or absence of VA mycorrhiza of the tropical species examined cannot be explained as yet. Apart from physiological processes in higher plant or fungus, changes in solubility or absorption on soil particles of PO_4 play a role, as experiments with other phosphorus compounds demonstrated.

Effect of water regime

With *E. odoratum* and *T. erecta*, Sieverding (1975) found that the amount of water used to produce 1 g of dry matter in a soil/sand mixture fertilized with $Ca_5(PO_4)_3OH$ was much smaller in mycorrhizal than in non-mycorrhizal plants. The plants generally used more water when $Ca_5(PO_4)_3OH$ was applied than with $Ca(H_2PO_4)_2 \cdot H_2O$. With $Ca_5(PO_4)_3OH$, the mycorrhizal effect was greatest in the drier soil. This effect need not be explained by improved uptake or transport of water; it may simply be due to the better utilization of water by plants not suffering from phosphorus deficiency.

In other experiments with *S. bicolor*, *G. abyssinica*, and *E. odoratum* in the soil/sand mixture and in an acidic sandy soil, Sieverding (1979) found a similar trend to improved water utilization by mycorrhizal plants receiving insoluble $Ca_5(PO_4)_3OH$, particularly at 10 per cent available water in the soil/sand mixture where the mycorrhiza was better developed than at the higher water regimes. This was not observed in the acidic sandy soil, indicating that the response of mycorrhiza to water supply is influenced by the soil type. Here, as with other factors, we are only at the beginning of understanding the complex system of soil/mycorrhiza/plant.

References

Graw, D. (1978). Der Einfluss der vesikulär-arbuskulären Mykorrhiza auf das Wachstum tropischer und subtropischer Pflanzen unter Berücksichtigung von Wirtsspezifität und pH-Wert des Bodens. Dissertation, Göttingen.

Nyabyenda, P. (1977). Einfluss der Bodentemperatur und organischer Stoffe im Boden auf die Wirkung der vesikulär-arbuskulären Mykorrhiza. Dissertation, Göttingen.

Sieverding, E. (1975). Einfluss von symbiontischen Pilzen auf Wachstum und Nährstoffaufnahme von *Eupatorium odoratum* L. und *Tagetes erecta* L. bei verschiedenen Wasserregimen und Phosphatformen im Boden. Diplom Thesis, Institut für Tropischen und Subtropischen Pflanzenbau, Göttingen.

—— (1979). Influence of water regime and soil temperature on the efficiency of VA mycorrhiza. Dissertation, Göttingen.

24 Effect of endomycorrhizae on growth and phosphorus uptake of *Agrostis* in a pot experiment

J. Pichot and B. Truong

Abstract

The favourable effect of endomycorrhizae on the growth of plants has been widely demonstrated, e.g. Mosse and Phillips (1971) and Khan (1972 and 1975) on maize and wheat under field conditions. The generally accepted explanation is an exploration of the soil volume by external associated hyphae beyond the immediate zone of depletion around the roots. This additional ability to absorb nutrients is particularly interesting in phosphorus deficient soils with a high fixing capacity. The part played by the endophyte in solubilizing the less-soluble forms of phosphorus is not quite clear. This study was undertaken to determine the efficiency of an indigenous population of endomycorrhizae from Upper Volta in an oxisol, with or without rock phosphate fertilizer, and to appreciate the modification of soil isotopically exchangeable phosphate by the endophyte in a pot culture.

Materials and method

Soil. The sample is an humic oxisol derived from basaltic alluvion, collected at Ambohimandroso (Madagascar), deficient in available phosphorus (14 p.p.m. TRUOG P) with high fixing capacity (860 p.p.m. P) and low pH (4.9 in water).

Endomycorrhizae. Young maize roots from Farako-Ba station (Upper Volta), well infected with mycelium and vesicles, were used.

Phosphate. A newly discovered rock phosphate deposit at Kodjari (Upper Volta), containing 13.16 per cent phosphorus and 31.67 per cent calcium, passed through a 0.1 mm sieve.

Pot experiments. (Five treatments and five replications)
  (i) Control.
  (ii) Soil fertilized with rock phosphate.
  (iii) Soil inoculated with fresh roots.

(iv) Soil + roots + phosphate.
(v) Soil + steamed roots in order to detect the effect of organic matter.

Rock phosphate was applied at the rate of 100 p.p.m. phosphorus. Fresh roots were cut into pieces 2 mm long and mixed with soil at a rate of 500 mg fresh weight or 106 mg dry weight/pot. Each polyethylene pot contained 100 g of soil, 0.1 mCi ^{32}P, and 5 mg of carrier phosphorus in the form of KH_2PO_4, other nutrients were added in sufficient quantities. 100 seeds of *Agrostis communis* were sown and covered with a small quantity of acid-washed fine sand. Plant leaves were harvested after the fifth and ninth weeks. At each harvest dry weight and phosphorus uptake were recorded and L values calculated according to the following formula:

$$L = {}^{32}P \text{ introduced} \left(\frac{{}^{31}P}{{}^{32}P} \text{ plant} \right) - \text{carrier } {}^{31}P.$$

After the experiment, root systems were collected, washed, and examined for mycorrhizal infection by clearing them for 30 min in 0.1 N KOH at 90 °C and 15 min in 5 per cent HCl and then staining in cotton blue in lactophenol.

Results and discussion

Averages of the five replicates are shown in Table 24.1. An analysis of variance between various treatments was performed.

Effect of rock phosphate.   The effect of rock phosphate was highly significant for all parameters recorded, and greater at the second harvest than at the first. For the total of the cuts, the fertilizer has increased the dry weight by 32 per cent, the phosphorus uptake by 50 per cent, and the L values by 30 p.p.m. This efficiency is to be expected as the soil tested was severely deficient in phosphorus and acid, it indicated that direct utilization of this new rock phosphate deposit is possible.

Effect of endomycorrhizae.   Mycorrhizae were only highly significant at the first harvest and for only two parameters. The dry weight increased by 50 per cent and phosphorus uptake by 42 per cent. The treatment phosphate + endomycorrhizae gave the highest results, 72 per cent more than the control. It seemed that direct inoculation of infected roots produced rapid effects on the host plant.

The effect on L values was not clear. The increases were only 4 or 5 p.p.m., and statistically not significant. Further experiments are needed, especially to find out whether the effect of endophyte concerned organic and/or inorganic phosphorus, e.g. an easily exchangeable pool.

Table 24.1. Pot experiment results

Treatments	Dry weight (mg)			Phosphate uptake (µg)			L value (p.p.m.)	
	1st cut	2nd cut	Total	1st cut	2nd cut	Total	1st cut	2nd cut
Control	212	379	591	381	562	943	31	31
Rock phosphate	273	508	781	559	858	1417	52	68
Endomycorrhizae	318	400	718	541	584	1125	34	35
Phosphate + myco.	365	501	866	659	833	1492	56	73
Steamed myco.	202	422	624	371	645	1016	32	32
Variation coefficient (%)	6.89	5.59	5.09	9.17	7.71	6.48	5.60	4.40
P effect without myco.	**	**	**	**	**	**	**	**
P effect with myco.	**	**	**	**	**	**	**	**
Myco. effect without P	**	NS	**	**	NS	**	NS	NS
Myco. effect with P	**	NS	**	**	NS	NS	NS	NS
P × myco. interaction	NS	NS	NS	NS	NS	NS	NS	NS

NS: non significant; * = significant (0.05 level); ** = significant (0.01 level).

Root examination.   Macroscopic examination of roots from all treatments showed normal mycelium, characteristically coloured with Cotton Blue; the infection was general in both indigenous and inoculated populations. Semiquantitative and detailed observations showed the following differences: the infection was sporadic in the control and phosphate treatments (about 10 per cent of the roots), the mycelia were long, granulous, less ramified, the wall net was pliant. Root hairs were abundant. The infection was continuous in the treatment with inoculum (about 25 per cent of roots); besides the long and flexible mycelium described previously, there were many others, e.g. bent, more ramified, segmented, and containing vesicles. Root hairs were scarce.

References

Khan, A. G. (1972). The effect of vesicular arbuscular mycorrhizal associations on growth of cereals. I. Effects on maize growth. *New Phytol.* 71, 613–19.

—— (1975). The effect of vesicular arbuscular mycorrhizal associations on growth of cereals. II. Effects on wheat growth. *Ann. appl. Biol.* 80, 27–36.

Mosse, B. and Phillips, J. M. (1971). The influence of phosphate and other nutrients on the development of vesicular-arbuscular mycorrhiza in culture. *J. gen. Microbiol.* 69, 157–68.

Sanders, F. E. and Tinker, P. B. (1971). Mechanism of absorption of phosphate from soil by *Endogone* mycorrhizas. *Nature, Lond.* 223, 278.

PART V

Mycorrhiza in tropical agriculture

25 Mycorrhiza in agricultural plants

B. Mosse and D. S. Hayman

General principles

Most agricultural plants form vesicular-arbuscular (VA) mycorrhiza. Potential hosts occur in the economically important families of Gramineae (e.g. maize and wheat) and Leguminoseae (e.g. soyabean and clover); many tropical plantation crops, e.g. coffee, tea, rubber, cocoa, oil palm, citrus, many tropical timber trees, also walnuts and all temperate fruit trees have VA mycorrhiza. Except for sporadic infections (Hirrel, Mehraveran, and Gerdemann 1978; Ocampo, Martin, and Hayman 1979) Cruciferae and Chenopodiaceae are usually non-mycorrhizal. Most temperate timber trees have ectomycorrhiza.

Probably the most important feature of VA mycorrhiza is the restricted spread of the fungus within the root and its extensive penetration of the soil beyond the zone of root exploration. VA infection causes little change in external appearance of the root. It is confined to the primary cortex and consists of intercellular distributive hyphae, finely branched intra-cellular structures, the arbuscules, and spore-like swellings, the vesicles, usually at the hyphal tip. Root-based and soil-based mycelium are directly connected by substantial hyphae (up to 15 μm in diameter) usually without septa. The soil-based mycelium often carries fine, thin-walled tufts of absorbing hyphae that develop septa as they lose their contents and become functionless. It may also carry large (100–600 μm) resting spores with characteristic diagnostic features that allow identification of some of the fungal species involved.

VA endophytes are obligate symbionts and can only multiply and spread in association with a host plant. They have very little host specificity. The spores can survive in the soil or laboratory for several years. Virtually all soils contain some VA endophytes but inoculum density and fungal species differ. Surveys of resting spores can give some indication of endophyte populations but spore numbers are not always a reliable index of soil infectivity or root infection because spore formation depends not only on inherent species characteristics but also on soil–fungus interactions; also some endophytes are non-sporing. The distribution of many endophyte species is world-wide but localized distribution patterns can be very variable and are usually unpredictable. Some species are restricted to tropical soils. Spore populations can change rapidly when virgin soil is brought into cultivation (Schenck and Kinloch 1976) because native endophytes are often sensitive to added fertilizers (Mosse 1977a; Powell and Daniel 1978a). Other, less sensitive species, then become established or can profitably be introduced. Hayman (1975) has reviewed changes in spore populations brought about by crop successions, fertilizer treatments, and effects of soil type (see also Kruckelmann 1975). Temperature and soil pH are other determining factors. The ability of a particular endophyte species to establish and compete with the indigenous microflora and endophyte population in a particular site can as yet only be determined experimentally. The general principles of VA mycorrhizal infection have recently been reviewed by Gerdemann (1975) and Hayman (1978).

Effects of VA mycorrhiza on plant growth: mode of action

Nutrient uptake

The main agronomic significance of VA mycorrhiza lies in their ability to supply the plant with extra nutrients, particularly those that, like phosphorus and zinc, diffuse so slowly in the soil that depletion zones develop in the immediate vicinity of the root. The interconnected network of external hyphae acts as additional catchment and absorbing surface in the soil beyond the depletion zone that would otherwise remain inaccessible to the plant roots. The uptake of ^{32}P 2.7 cm from the root surface has been demonstrated (Hattingh, Gray, and Gerdemann 1973), and many experiments have shown several-fold increases in growth and phosphorus uptake resulting from mycorrhizal infection. Phosphorus taken up by the fungus in the soil is translocated, probably by mass flow of polyphosphate granules, into the root-based mycelium and there released for plant use.

The extent to which the plant benefits from mycorrhizal association depends on its phosphorus requirement and inherent ability to forage for phosphate, on soil reserves of available phosphorus, and on the fungal species. Plants with extensive fine roots and numerous long root hairs, e.g. many grasses, tend to be good foragers and less dependent on mycorrhizal infection than those with short fleshy roots lacking root hairs, e.g. citrus and onions. Nevertheless even the former can benefit from mycorrhiza in very phosphorus-deficient soils. Slow-growing species may also be less mycorrhiza-dependent. An adequate P supply is particularly important for legumes because symbiotic nitrogen fixation has a high phosphorus requirement and roots and nodules compete for phosphate. Nodules may have the first call on limited supplies (Smith and Daft 1977). In competition with grasses legumes are poor foragers for phosphate. They can only maintain themselves in pastures with the advantages conferred by symbiotic nitrogen fixation which in turn often depends on phosphorus uptake through mycorrhiza. A notable exception are some *Lupinus* species (Asai 1944; Trinick 1977).

Because mycorrhizal and non-mycorrhizal plants use the same sources of soil phosphate the available phosphorus resources of the soil largely determine inoculation responses. If such resources are extremely low, as in many tropical soils, mycorrhizal infection may cause a 50 per cent increase in dry matter but yet not give an economic yield. Unless reserves of available phosphorus are maintained or augmented by adding fertilizer the advantages of an efficient mycorrhizal system will be short-lived. Increased utilization of soluble and sparingly soluble phosphorus fertilizer by mycorrhizal plants has been shown in many experiments.

Finally the benefits of mycorrhizal association depend on the endophyte. Endophyte species differ in their effects on plant growth and their performance depends to some extent on interactions between fungus and soil. Although some endophytes appear to be consistently poor, the best endophyte in one soil may be surpassed by an apparently inferior one in another soil (Fig. 25.1). All endophytes appear to use the same sources of soil phosphate. Their performance is not always related to infection levels, or to the number of connecting links between soil and root-based mycelium (Powell 1977a). Several factors may control endophyte efficiency including the degree of pathogenicity which could vary with host species. When the host plant is adequately provided with phosphorus mycorrhizal infection can reduce growth (Sparling and Tinker 1978; Johnson 1976). This has, however, not yet been demonstrated with plants inoculated and grown in natural, as opposed to sterilized soils. Sometimes such growth depressions are temporary, immediately following infection; at other times they have become reversed when plants were kept until the soil became very depleted (Cooper 1975).

The VA mycorrhizal system then, involves three components (Fig. 25.2), the plant, the soil, and the fungus. Interactions between the three determine the mycorrhizal effect on plant growth. Although mycorrhiza may depress growth in very fertile soils infection more usually dies out under such conditions, because high internal nutrient levels of both nitrogen and phosphorus can make the plant immune to infection (Mosse 1973; Sanders 1975; Menge, Steirle, Bagyaraj, Johnson, and Leonard 1978). In soils of low and moderate phosphorus content infection will lead to growth increases particularly in the more mycorrhiza-dependent plant species. The endophyte species can affect the size of such increases. To fulfil its potential the endophyte must be well adapted to the environmental conditions.

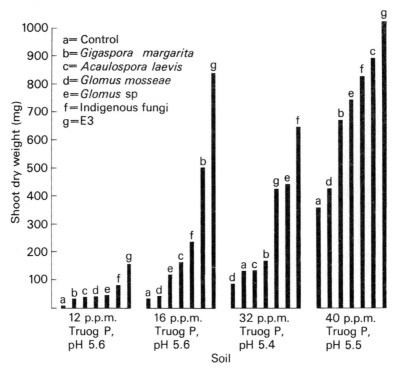

Fig. 25.1. Growth of *Trifolium repens* inoculated with six different endophytes in each of four sterilized soils (after Powell 1977a).

The effects of inoculation will therefore vary. They can be expressed as kg P/ha required to produce similar growth increments. Some figures taken from the literature are 30 kg P/ha for ryegrass, 56 kg P/ha for maize, wheat, and barley, 160 kg P/ha for cassava, 176 kg P/ha for soyabean, and 556 kg P/ha for sour orange.

In many tropical soils sulphur deficiency can reduce plant growth. Cooper and Tinker (1978) showed that sulphur and zinc could be taken up by fungal hyphae and translocated to the host but mean fluxes of phosphorus, sulphur, and zinc were in the ratio of 50:8:1. Nevertheless, zinc deficiency symptoms in peaches were cured by mycorrhizal inoculation (La Rue, McClellan, and Peacock 1975; Table 25.1). Mycorrhizal onions took up sulphur applied 8 cm from the root surface but non-mycorrhizal onions did not; they took up only one-eighth as much as mycorrhizal ones when the sulphur source was 5 cm from the root surface (Rhodes and Gerdemann 1978).

More detailed reviews of the relevance of VA mycorrhiza to plant

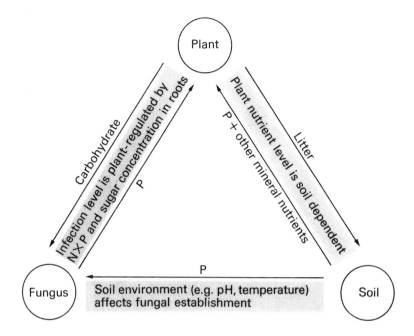

Fig. 25.2. Some factors involved in the functioning of mycorrhizal systems.

nutrition in agriculture are those of Tinker (1979), Mosse and Tinker (1980), and Mosse (1977b, 1979).

Non-nutritional effects

Mycorrhiza also affect disease resistance, soil aggregation, and transplanting. Ability to transplant is particularly important for tree crops normally raised in nurseries. Many are nowadays routinely fumigated or treated with nematicides. This can reduce or eliminate the soil population of VA endophytes and lead to non-mycorrhizal nursery stock being transplanted to permanent sites. Containerized ornamentals may be included in this group. Often such stock is slow to establish and may even die (Bryan and Kormanik 1977; Menge, Davis, Johnson, and Zentmyer 1978), especially under adverse environmental conditions like those in coal spoils (Daft and Hacskaylo 1976, 1977; Khan 1978), dunes, and eroded and degraded sites. Sometimes death after transplanting is due to water stress; phosphorus deficient seedlings are known to be more sensitive to this condition (Atkinson and Davison 1972). The greater internal resistance to water flow in non-mycorrhizal plants was also a phosphorus effect (Safir, Boyer, and Gerdemann 1972). Other, as yet undetermined reasons may contribute to the better survival of mycorrhizal plants under adverse conditions.

VA mycorrhizal fungi are better able than many other soil fungi to bind soil into semi-stable aggregates (Sutton and Sheppard 1976). Clough and Sutton (1978) attributed this to the production of gum-like substances, possibly polysaccharides, formed by the fungi or associated bacteria. This could make mycorrhizal plants particularly useful for the rehabilitation of eroded sites.

Soil-based fungal diseases fall into three main groups: feeder root diseases, collar rots, and vascular wilt diseases. Mycorrhizal infection can decrease the pathogenic effects of certain feeder root disease fungi by physically excluding them from already occupied sites (Becker 1976; Stewart and Pfleger 1977), although the VA fungi form no barrier corresponding to the ectomycorrhizal sheath. More often their effect is physiological, the increased vigour of plants with mycorrhiza making them less sensitive to stress resulting from root death (e.g. Schönbeck and Dehne 1977). Increased chlorophyll content and resistance to cell leakage are suggested reasons for mycorrhizal plants being less affected than non-mycorrhizal ones by vascular wilt diseases (Dehne and Schönbeck 1975). Although not known to produce antibiotics, as found in ectomycorrhiza, tobacco roots with VA mycorrhiza contained more arginine which inhibited infection and sporulation by a root pathogen (Baltruschat and Schönbeck 1975). Conversely, *Phytophthora cinnamomi* in avocado roots inhibited mycorrhizal infection and spore production and prevented mycorrhizal enhancement of phosphorus uptake in one experiment (Davis, Menge, and Zentmyer 1978) but had no effect in another (Mataré and Hattingh 1978). Other root diseases seem to co-exist with VA mycorrhizal infections (O'Brien and McNaughton 1928; Timmer and Leyden 1978b; Davis *et al.* 1978). According to Schönbeck (1979) susceptibility to some leaf diseases is increased in plants with VA mycorrhiza. Plant-pathogenic nematodes also show a wide range of interactions with VA mycorrhiza. Frequently there is an inverse relationship between the populations of nematodes and VA fungi. Biological control of plant diseases by VA mycorrhiza may be possible in some circumstances but as yet there is insufficient evidence of this. A comprehensive review of the literature pertaining to the interaction of VA mycorrhiza with plant pathogens was made by Schenck and Kellam (1978).

Field inoculation

One objective of mycorrhizal research is its application to crop improvement and better utilization of applied fertilizer in the field. Pot experiments in sterilized soils have laid the foundation to an understanding of some of the principles underlying the functioning of VA mycorrhiza but they are poor indicators of field responses. For instance, inoculation increased growth of clover in a Welsh hill-side pasture but there was little response in a parallel pot experiment (Hayman and Mosse 1978, 1979). In a New Zealand hillside pasture inoculation responses were better or worse than those in the glass-house according to soil and fungal species (Powell and Daniel 1978b). Maize responded well to inoculation in a pot experiment but not in the same soil in the field (Islam 1976). Tussock grassland species were more responsive to inoculation at high altitudes than in a glass-house (Crush 1973). Probably no single explanation accounts for such varied discrepancies but it is evident that only field trials can evaluate the significance of inoculation with selected mycorrhizal fungi for agricultural crops.

The small scale trials of VA mycorrhizal inoculation so far reported in unsterilized soils are summarized in Table 25.1. They involve grain crops, tropical and temperate forage and grain legumes, potatoes, and onions in Britain, New Zealand, Nigeria, India, and Pakistan. Similar experiments are in progress in Australia, Brazil, and Hawaii. The figures indicate a wide range of results. Lack of response may be genuine but can also indicate that factors other than phosphorus uptake controlled growth; a plus phosphate treatment should therefore always be included in the design. It may also indicate that the introduced endo-phyte did not become established. If percentage root infection in the inoculated plants is higher than in the non-inoculated, this is usually taken as some indication of successful establishment. Confirmation by anatomical detail of the infection in the root and spore production of the introduced endophyte in the soil are other useful criteria. Finally inoculum can be transferred from one plot to another during necessary cultivation and precautions should be taken against this by adequate spacing between plots. Even if inoculum is scarce there is no limitation on size and replication of control (uninoculated) plots.

Most of the tests in Table 25.1 show positive inoculation responses. Two general criticisms can be made of the experiments. Some of the soils chosen were of unusually low infectivity after prolonged fallow-ing and/or contained so little available phosphorus that they would not normally be in agricultural use. Secondly an inoculation tech-nique like the outplanting of pre-inoculated seedlings, is impracticable for field-sown annuals and gives an initial advantage to the inocu-lated plants compared to the uninoculated that pick up infection more slowly from indigenous endophytes in the soil.

Phosphate status of sites

Not all the soils chosen for inoculation experiments were extremely phosphorus deficient and results indicate that inoculation responses

Table 25.1. Published field experiments on inoculation with VA mycorrhiza

Crop	VA species‡	Soil status	Inoculation method	Growth increases	Country	Reference
Maize	Glomus mosseae	U	a	×2.3	Pakistan	Khan (1972)
Wheat	G. mosseae	U	a	×3	Pakistan	Khan (1975)
Barley	G. mosseae	U	a	×4	Pakistan	Saif and Khan (1977)
Barley	G. caledonius	U	c	×1.3 ⎫	UK	Owusu-Bennoah and Mosse (1979)
Onion	G. caledonius	U	c	×6 ⎬		
Lucerne	G. caledonius	U	c	×4 ⎭		
Maize	G. mosseae	⎰U⎱ ⎱F⎰	b b	×1.3 (×1.6)* ×1.3 (×1.3)*	Nigeria	Islam (1977)
Cowpea	G. mosseae	U	a b	×1.3 (×1.3)* ×1.3 (×1.5)*		
Soyabean	G. fasciculatus	U	c	0	India	Bagyaraj, Manjunath, and Patil (1979)
Potatoes	Indigenous	U	b	×1.2	UK	Black and Tinker (1977)
White clover	G. fasciculatus + Indigenous	U	a	0 to ×2.7	New Zealand	Powell (1977b)
White clover	G. fasciculatus + G. mosseae	U	a	0 to ×3	UK	Hayman and Mosse (1979)
White clover	G. tenuis + Gigaspora margarita	U	d	×1.8	New Zealand	Powell (1979)
White clover	Several	U	a	×1.1 to 2.2	New Zealand	Powell and Daniel (1978b)
Citrus	Glomus mosseae	F	c	0 to ×2.5	US (Illinois)	Kleinschmidt and Gerdemann (1972)

Crop	Endophyte	Soil		Response	Location	Reference
Citrus	Gigaspora calospora	U F		0	US (Florida)	Schenck and Tucker (1974)
	Glomus macrocarpus var. geosporus	U F		X2		
Citrus	G. fasciculatus	F	c d	X1.7 & 4	US (California)	Hattingh and Gerdemann (1975)
Citrus	G. fasciculatus	F	c	0	US (Texas)	Timmer and Leyden (1978b)
Peach	G. fasciculatus	F	c	X1.8	US (California)	La Rue et al. (1975)
Sweetgum	G. mosseae	F	c	X52	US (Georgia)	Bryan and Kormanik (1977)
	Indigenous			X84		
Soyabean	G. macrocarpus var. geosporus	F	c	0 & X1.3	US (N. Carolina)	Ross and Harper (1970)
Soyabean	Gigaspora calospora	F	a	0 & X1.5	US (Florida)	Schenck and Hinson (1973)
Cassava	Indigenous	UF	—	X3†	US (Hawaii)	Vander Zaag, Fox, de la Pena, and Yost (1979)

U = unsterile soil; F = fumigated soil (in some experiments control plants were strongly mycorrhizal when harvested). a = pre-inoculated transplants; b = highly infective field soil placed in seed furrows; c = soil + roots from pot cultures (stock plants) placed in seed furrows; d = inoculum pelleted onto seed.

*Grain yield.

†Per cent phosphorus.

‡Species names of VA endophytes are those given by the authors. Isolates with the same name are not necessarily identical.

can often be better in moderately fertile soils. Adding 90 kg P/ha (as basic slag) increased clover response to inoculation by 118 per cent (Hayman and Mosse 1979). Inoculation responses of onion and lucerne, but not of barley, were greater in a field soil with 14 p.p.m. available phosphorus (Olsen, Cole, Watanabe, and Dean 1954) than with 10 p.p.m. (Owusu-Bennoah and Mosse 1979). Cowpeas responded well to inoculation in a moderately fertile soil depleted by one crop of maize (Islam 1977). Inoculation responses were suppressed in maize, wheat, and barley by 56 kg P/ha (triplesuperphosphate) (Khan 1972, 1975; Saif and Khan 1977) and in potatoes by 82 kg P/ha (Black and Tinker 1977). These additions also reduced mycorrhizal infection thus lowering inoculum potential for a subsequent crop that might then require more fertilizer phosphorus for satisfactory growth. Under some conditions fertilizers with lower phosphorus availability might have practical advantages over more available forms if they were less inhibitory to mycorrhizal infection and could be made more accessible to the plant by an efficient mycorrhizal system. This applies particularly in tropical soils with high 'phosphorus fixing' capacity where it is difficult to add sufficient phosphorus fertilizer to correct the naturally low phosphate status. In such situations the cost and inaccessibility of manufactured fertilizers can also make the use of naturally occurring rock phosphate very attractive. In pot experiments mycorrhizal inoculation increased utilization not only of soluble phosphates like monocalcium and superphosphate (Mosse 1973; Menge, Johnson, and Platt 1978; Powell and Daniel 1978a) but also of less soluble forms like rock phosphate (Mosse 1977a), bonemeal (Daft and Nicolson 1966), and even hydroxyapatite and aluminium and iron phosphates (Nyabyenda 1977). Temperature markedly affected responses to the latter. The high temperature of tropical soils would increase fungal activity provided the tolerance range of the fungus was not exceeded.

Even in more sophisticated agricultural systems reinoculation after fumigation may be of practical interest. Relevant studies, also summarized in Table 25.1, have been confined mainly to fumigated citrus nurseries in the United States where methyl bromide and chloropicrin are widely used to control pathogens and nematodes. These fumigants also eliminate or greatly reduce VA endophytes. Although very high phosphorus inputs (above 500 kg P/ha) can compensate for this, the non-mycorrhizal plants may then become copper deficient (Timmer and Leyden 1978a). Cotton has sometimes developed zinc deficiency as a result of removing VA endophytes by fumigation (Wilhelm, George, and Pendery 1967). Not all nematicides suppress VA endophytes, and mycorrhizal infection remains very high in pineapple roots in Hawaiian plantations that are regularly treated with nematicides. Strawberries in soil treated with DD also remained strongly mycorrhizal and infection in barley was sometimes increased by aldicarb (Ocampo and Hayman 1979).

Inoculation techniques

Three techniques have been used in field experiments:

(i) Pre-inoculation of transplanted seedlings. Inoculum is placed in the seed pan and seedlings are transplanted to the experimental site when adequately infected.

(ii) Inoculation at the experimental site by placing infected soil and/or roots below the seed.

(iii) Pelleting seed with infected soil.

Pre-inoculation has been used extensively because it was thought to provide the best chance for establishing a selected endophyte in competition with those already in the soil. Apart from the possibility that there may already be a nutritional difference between mycorrhizal and non-mycorrhizal seedlings at the transplanting stage, other criticisms of the technique are impracticability and delayed infection of the control plants. While obviously valid for field-sown annuals the argument of impracticability applies much less to nursery stock. Furthermore, the objective of these early experiments was to evaluate the potential of field inoculation rather than its practicability.

Placing inoculum below the seed proved as effective as pre-inoculation in experiments with cowpea under Nigerian conditions; pre-inoculated and seed-inoculated plants were of equal size after 48–88 days (Islam 1977). A 10 g inoculum placed below the seed became well established (Owusu-Bennoah and Mosse 1979) and spores of the introduced fungus were found 22 cm from the point of inoculation after 13 weeks (Warner 1980). Pellets of only 1 cm diameter made from a mixture of soil and seed in the ratio of 40:1 were successfully used by Powell (1979). Methyl cellosolve (Kleinschmidt and Gerdemann 1972) or clay (Hall 1979) have been used for pelleting and other protective carrier materials such as peat or lignite could be tried. Incorporating rock phosphate into such pellets has been considered. The objective of all such methods is the strategic placement of inoculum near the emerging seedling roots so that infection by selected endophytes is favoured and develops sooner than in un-inoculated seedlings dependent upon the more dispersed propagules of naturally occurring endophytes. An adequate nutrient supply during early growth is often reflected in increased crop yields, especially for short-lived annuals. For trees the advantages gained during the first year are generally maintained subsequently.

More field experiments are needed before a valid judgement can be made of the potential value of field inoculation. For the foreseeable future the selection of suitable endophytes will remain largely empirical. The mycorrhiza dependence of some host species like onion, citrus, cassava, and cowpea is so great that they are likely to respond to inoculation even in fertile soils. Recent findings suggest that varieties within a species can apparently differ in this respect (Crush 1978; Hall 1978; Kormanik, Bryan, and Schultz 1977; Mosse and Thompson 1980). Pre-knowledge of the most responsive varieties and species would help to rationalize any inoculation programme.

Practical considerations

Producing sufficient inoculum is likely to be the main obstacle to large scale inoculation, but problems of storage and inoculation technique also require solution. Three types of inocula are possible: pure fungal cultures, infected roots, and infested soil.

Fungal cultures

Pure cultures of VA endophytes cannot so far be grown axenically on synthetic media. They would have advantages for quality control of inoculum and freedom from contamination by other endophytes and pathogens, but they are not necessarily more infective or better able to maintain viability during storage than infected roots or infested soil. Resting spores of VA endophytes are probably the most resistant structures but they are large, limited in number, and not easily detached from the mycelium and would therefore be difficult to harvest in quantity. Hall (1976) found that an inoculum of root segments caused more rapid growth stimulation than spores. Powell (1976) showed that hyphae from germinating spores had a pre-infection stage during which they ramified over the root surface before infection occurred whereas hyphae arising from infected root pieces penetrated more directly. Very recently a VA endophyte was found producing very many, easily harvested spores in epigeous sporocarps (Daniels and Trappe, personal communication).

Infected roots

Infected roots can be raised in sterilized soil, sand, or liquid culture. If considerations of storage require roots to be freed from the supporting medium, sand or liquid culture are preferable. Heavy infections can develop in sand culture (Daft and Nicolson 1966) and in sand and vermiculite (Hepper, personal communication) particularly if bonemeal is used as a phosphate source. A technique of growing infected roots in nutrient flow culture has been pioneered by Mosse and Thompson (1980). The composition of the nutrient solution affected the type of infection and the amount of external mycelium produced. Sporocarps and resting spores were also formed. As little as 0.08 g of root inoculum produced 50 per cent infection in maize after six weeks. The inoculum remained infective after air-drying. Air-dried sievings, consisting of soil mycelium and infected root fragments also retained infectivity after three years (Mosse, unpublished). Such material might be incorporated into multi-seeded pellets with an inert carrier such as charcoal or peat and a fertilizer base.

Successful lyophilization of infected roots has been claimed by Jackson, Franklin, and Miller (1972) but in more extensive trials lyophilization led to markedly reduced infectivity (Crush and Pattison 1975). Resting spores survived best.

Infested soil

For many years forest trees were inoculated with ectomycorrhiza by simple transfer of some surface soil from an infected site to the planting hole of the new seedling. Inoculum and disease control were insufficient but the method could be improved and adapted to produce soil infested with VA endophytes in large containers or even on a semi-field scale. Soil could be fumigated or sterilized with formalin, planted with a suitable host plant, and re-inoculated with material raised in properly controlled pot cultures. Eventually soil and roots would be harvested, stored, and used as inoculum. Infested soil stored in a cool place in plastic bags retains its infectivity for at least six months and generally longer. Menge, Lembright, and Johnson (1977) proposed a scheme for multiplying inoculum, originating from carefully monitored, spore-inoculated mother plants in sterilized soil in pots, on an intermediate host, e.g. Sudan grass, grown in seed boxes or large containers. The intermediate host should not be susceptible to the same pests and diseases as the species for which the inoculum is intended. Non-toxicity of root residues of the multiplying host is important. From this point of view Sudan grass may not be a good choice.

Inoculum placement and quantities

The inoculation of crops like coffee, citrus, cassava, or tobacco, raised in restricted areas of nursery beds or seed boxes, presents fewer problems than that of annuals like wheat, maize, or soyabean seeded over large areas. Furthermore, inoculum placement is probably less critical in the generally sterilized nursery soil than in unsterile field soil with competing indigenous endophytes. For seeded crops the inoculum can be attached to the seed in some kind of pellet or it can be drilled with the seed either dry or as a slurry (Witty and Hayman 1978). Attachment to the seed has no particular advantage as the inoculum has a very limited growth range until it finds a host root, and neither resting nor vegetative spores are easily disseminated in the soil. Inoculum that remains attached to the seed coat will be of little use if active root growth has spread away from the seed. Placement some centimetres below the seed may therefore be better.

In two field experiments a 10 g inoculum of strongly infested soil and roots placed below each planting hole led to good inoculum establishment and improved growth (Islam 1977; Owusu-Bennoah and Mosse 1979). Assuming an effective spread of 10 cm from the point of inoculation, 50 g of inoculum would be required per metre row. With a 20 cm distance between rows 2500 kg/ha would be needed. Grain crops with a sowing rate of 200 kg/ha, pelleted with inoculum at the rate of 1:40, would require 8000 kg/ha of infested soil. Such quantities are impracticable to produce, handle, and transport. While pelleted seed may be feasible for legumes sown in pastures, some less bulky inoculum would be needed for annuals. Air-dried root inoculum might meet this requirement.

The economics of field inoculation with VA endophytes cannot be assessed at present. Its value in terms of increased crop, better establishment after transplanting, and more economic use of applied fertilizer has to be offset against cost of inoculum production, treatment, and application. Because nearly all agricultural plants have mycorrhiza the potential market is large and because there is little host specificity the number of endophyte species needed for distribution would be small. Persistence of the effects of inoculation and spread of inoculum in the field will also affect the calculation. Growth effects on clover in an upland pasture were greater in the second than in the first year after inoculation (Hayman and Mosse 1979). The cost of inoculation might, therefore, be distributed over several years. Even if inoculation proves too costly, crop successions and cultural practices might be manipulated so as to stimulate symbiotic association with the naturally occurring endophytes.

References

Asai, T. (1944). Über die Mykorrhizenbildung der Leguminosen-Pflanzen. *Jap. J. Bot.* 13, 463–85.

Atkinson, D. and Davison, A. W. (1972). The effects of phosphorus deficiency on water content and response to drought. *New Phytol.* 72, 307–13.

Bagyaraj, D. J., Manjunath, A., and Patil, R. B. (1979). Interaction between a vesicular-arbuscular mycorrhiza and *Rhizobium* and their effects on soybean in the field. *New Phytol.* 82, 141–5.

Baltruschat, H. and Schönbeck, F. (1975). Untersuchungen über den Einfluss der endotrophen Mycorrhiza auf den Befall von Tabak mit *Thielaviopsis basicola. Phytopath. Z.* 84, 172–88.

Becker, W. N. (1976). Quantification of onion vesicular-arbuscular mycorrhizae and their resistance to *Pyrenochaeta terrestris*. Ph.D. thesis, University of Illinois.

Black, R. L. B. and Tinker, P. B. (1977). Interaction between effects of vesicular-arbuscular mycorrhiza and fertilizer phosphorus on yields of potatoes in the field. *Nature, Lond.* 267, 510-11.

Bryan, W. C. and Kormanik, P. P. (1977). Mycorrhizae benefit survival and growth of sweetgum seedlings in the nursery. *Sth. J. appl. For.* 1, 21-3.

Clough, K. S. and Sutton, J. C. (1978). Direct observation of fungal aggregates in sand dune soil. *Can. J. Microbiol.* 24, 333-5.

Cooper, K. M. (1975). Growth responses to the formation of endotrophic mycorrhizas in *Solanum, Leptospermum*, and New Zealand ferns. In *Endomycorrhizas* (eds. F. E. Sanders, B. Mosse, and P. B. Tinker) pp. 391-407. Academic Press, London.

— and Tinker, P. B. (1978). Translocation and transfer of nutrients in vesicular-arbuscular mycorrhizas. II Uptake and translocation of phosphorus, zinc and sulphur. *New Phytol.* 81, 43-52.

Crush, J. R. (1973). Significance of mycorrhizas in tussock grassland in Otago, New Zealand. *N.Z. Jl. Bot.* 11, 645-60.

— (1978). Changes in effectiveness of soil endomycorrhizal fungal populations during pasture development. *N.Z. Jl. agric. Res.* 21, 683-5.

— and Pattison, A. C. (1975). Preliminary results on the production of vesicular-arbuscular mycorrhizal inoculum by freeze drying. In *Endomycorrhizas* (eds. F. E. Sanders, B. Mosse, and P. B. Tinker) pp. 485-93. Academic Press, London.

Daft, M. J. and Hacskaylo, E. (1976). Arbuscular mycorrhizas in the anthracite and bituminous coal wastes of Pennsylvania. *J. appl. Ecol.* 13, 523-31.

— — (1977). Growth of endomycorrhizal and non-mycorrhizal red maple seedlings in sand and anthracite soil. *Forest Sci.* 23, 207-16.

— and Nicolson, T. H. (1966). Effect of *Endogone* mycorrhiza on plant growth. *New Phytol.* 65, 343-50.

Davis, R. M., Menge, J. A., and Zentmyer, G. A. (1978). The influence of vesicular-arbuscular mycorrhizae on *Phytophthora* root rot of three crop plants. *Phytopathology* 68, 1614-17.

Dehne, H. W. and Schönbeck, F. (1975). Untersuchungen über den Einfluss der endotrophen Mykorrhiza auf die Fusarium-Welke der Tomate. *Z. PflKrankh. Pflschutz* 82, 630-2.

Gerdemann, J. W. (1975). Vesicular-arbuscular mycorrhiza. In *The development and function of roots* (eds. J. G. Torrey and D. T. Clarkson) pp. 575-91. Academic Press, London.

Hall, I. R. (1976). Response of *Coprosma robusta* to different forms of endomycorrhizal inoculum. *Trans. Br. mycol. Soc.* 67, 409-11.

— (1978). Effect of vesicular-arbuscular mycorrhizas on two varieties of maize and one of sweetcorn. *N.Z. Jl. agric. Res.* 21, 517-19.

— (1979). Soil pellets as a method of introducing vesicular-arbuscular mycorrhizal fungi into soil. *Soil Biol. Biochem.* 11, 85-6.

Hattingh, M. J. and Gerdemann, J. W. (1975). Inoculation of Brazilian sour orange seed with an endomycorrhizal fungus. *Phytopathology* 65, 1013-16.

— Gray, L. E., and Gerdemann, J. W. (1973). Uptake and translocation of ^{32}P-labelled phosphate to onion roots by endomycorrhizal fungi. *Soil Sci.* 116, 383-7.

Hayman, D. S. (1975). The occurrence of mycorrhiza in crops as affected by soil fertility. In *Endomycorrhizas* (eds. F. E. Sanders, B. Mosse, and P. B. Tinker) pp. 495-509. Academic Press, London.

Hayman, D. S. (1978). Endomycorrhizae. In *Interactions between non-pathogenic soil microorganisms and plants* (eds. Y. R. Dommergues and S. W. Krupa) pp. 401–42. Elsevier, Amsterdam.

— and Mosse, B. (1978). Clover in grassland. Rothamsted Rep. for 1977, p. 241.

— — (1979). Improved growth of white clover in hill grasslands by mycorrhizal inoculation. *Ann. appl. Biol.* 93, 141–8.

Hirrel, M. C., Mehraveran, H., and Gerdemann, J. W. (1978). Vesicular-arbuscular mycorrhizae in the *Chenopodiaceae* and *Cruciferae*: do they occur? *Can. J. Bot.* 56, 2813–17.

Islam, R. (1976). Progress report on mycorrhiza research at the International Institute of Tropical Agriculture, Ibadan, Nigeria. IITA Internal Report.

— (1977). Effect of several *Endogone* spore types on the yield of *Vigna unguiculata*. International Institute of Tropical Agriculture, Ibadan, Nigeria. IITA Internal Report.

Jackson, N. E., Franklin, R. E., and Miller, R. H. (1972). Effects of VA mycorrhizae on growth and phosphorus content of three agronomic crops. *Soil Sci. Soc. Proc.* 36, 64–7.

Johnson, P. N. (1976). Effects of soil phosphate level and shade on plant growth and mycorrhizas. *N.Z. Jl. Bot.* 14, 333–40.

Khan, A. G. (1972). The effect of vesicular-arbuscular mycorrhizal associations on growth of cereals. I. Effects on maize growth. *New Phytol.* 71, 613–19.

— (1975). The effect of VA mycorrhizal associations on growth of cereals. II. Effects on wheat growth. *Ann. appl. Biol.* 80, 27–36.

— (1978). Vesicular-arbuscular mycorrhizas in plants colonizing black wastes from bituminous coal mining in the Illawara region of New South Wales. *New Phytol.* 81, 53–63.

Kleinschmidt, G. D. and Gerdemann, J. W. (1972). Stunting of citrus seedlings in fumigated nursery soils related to the absence of endomycorrhizae. *Phytopathology* 62, 1447–53.

Kormanik, P. P., Bryan, W. C., and Schultz, R. C. (1977). Influence of endomycorrhizae on growth of sweetgum seedlings from eight mother trees. *Forest Sci.* 23, 500–6.

Kruckelmann, H. W. (1975). Effects of fertilizers, soils, soil tillage and plant species on frequency of *Endogone* chlamydospores and mycorrhizal infection in arable soils. In *Endomycorrhizas* (eds. F. E. Sanders, B. Mosse, and P. B. Tinker) pp. 511–25. Academic Press, London.

La Rue, J. H., McClellan, W. D., and Peacock, W. L. (1975). Mycorrhizal fungi and peach nursery nutrition. *Calif. Agric.* 29, 7–8.

Mataré, R. and Hattingh, M. J. (1978). Effect of mycorrhizal status of avocado seedlings on root rot caused by *Phytophthora cinnamomi*. *Pl. Soil* 49, 433–5.

Menge, J. A., Johnson, E. L. V., and Platt, R. G. (1978). Mycorrhizal dependency of several citrus cultivars under three nutrient regimes. *New Phytol.* 81, 553–60.

— Lembright, H., and Johnson, E. L. V. (1977). Utilization of mycorrhizal fungi in citrus nurseries. *Proc. Int. Soc. Citriculture*, Vol. 1, pp. 129–32.

— Davis, R. M., Johnson, E. L. V., and Zentmyer, G. A. (1978). Mycorrhizal fungi increase growth and reduce transplant injury in avocado. *Calif. Agric.* 32, 6–7.

— Steirle, D., Bagyaraj, D. J., Johnson, E. L. V., and Leonard, R. T. (1978). Phosphorus concentrations in plants responsible for inhibition of mycorrhizal infection. *New Phytol.* 80, 575–8.

Mosse, B. (1973). Advances in the study of vesicular-arbuscular mycorrhiza. *A. Rev. Phytopathol.* 11, 171–96.

— (1977a). Plant growth responses to vesicular-arbuscular mycorrhiza. X. Responses of *Stylosanthes* and maize to inoculation in unsterile soils. *New Phytol.* 78, 277–88.

— (1977b). The role of mycorrhiza in legume nutrition on marginal soils. In *Exploiting the legume*–Rhizobium *symbiosis in tropical agriculture* (eds. J. M. Vincent, A. S. Whitney, and J. Bose) pp. 275–92. College of Tropical Agriculture, University of Hawaii. Misc. Publ. 145.

— (1979). *VA mycorrhiza research in tropical soils.* College of Tropical Agriculture, University of Hawaii. Misc. Publ.

— and Thompson, J. P. (1980). Nutrient film culture of vesicular-arbuscular mycorrhiza for mass production of inoculum. (In preparation.)

— and Tinker, P. B. (1980). Effects of mycorrhizas on nutrition of higher plants. *CRC handbook of nutrition and food.* (In press.)

Nyabyenda, P. (1977). Einfluss der Bodentemperatur und organischer Stoffe auf die Wirkung der vesikulär-arbuskulären Mykorrhiza. Dissertation, Göttingen.

O'Brien, D. G. and McNaughton, E. J. (1928). The endotrophic mycorrhiza of strawberries and its significance. Res. Bull. No. 1, W. Scotland Agric. Coll.

Ocampo, J. A. and Hayman, D. S. (1979). Effect of pesticides on VA mycorrhiza. Rothamsted Report for 1978, Part 1.

— Martin, J., and Hayman, D. S. (1979). Influence of plant interactions on vesicular-arbuscular mycorrhizal infections. I. Host and non-host plants grown together. *New Phytol.* 84, 27–35.

Olsen, S. R., Cole, C. V., Watanabe, F. S., and Dean, L. A. (1954). Estimation of available phosphorus in soils by extraction with sodium bicarbonate. *Circ. U.S. Dep. Agric.* 939.

Owusu-Bennoah, E. and Mosse, B. (1979). Plant growth responses to vesicular-arbuscular mycorrhiza. XI Field inoculation responses in barley, lucerne and onion. *New Phytol.* 83, 671–9.

Powell, C. L. (1976). Development of mycorrhizal infections from *Endogone* spores and infected root segments. *Trans. Br. mycol. Soc.* 66, 439–45.

— (1977a). Mycorrhizas in hill country soils. II. Effect of several mycorrhizal fungi on clover growth in sterilised soils. *N.Z. Jl. agric. Res.* 20, 59–62.

— (1977b). Mycorrhizas in hill country soils. III. Effect of inoculation on clover growth in unsterile soils. *N.Z. Jl. agric. Res.* 20, 343–8.

— (1979). Inoculation of white clover and ryegrass seed with mycorrhizal fungi. *New Phytol.* 83, 81–6.

— and Daniel, J. (1978a). Mycorrhizal fungi stimulate uptake of soluble and insoluble phosphate fertilizer from a phosphate-deficient soil. *New Phytol.* 80, 351–9.

— — (1978b). Growth of white clover in undisturbed soils after inoculation with efficient mycorrhizal fungi. *N.Z. Jl. agric. Res.* 21, 675–81.

Rhodes, L. H. and Gerdemann, J. W. (1978). Hyphal translocation and uptake of sulphur by vesicular-arbuscular mycorrhizae of onion. *Soil. Biol. Biochem.* 10, 355–60.

Ross, J. P. (1972). Influence of *Endogone* mycorrhiza on *Phytophthora* rot of soybean. *Phytopathology* 62, 896–7.

— and Harper, J. A. (1970). Effect of *Endogone* mycorrhiza on soybean yields. *Phytopathology* 60, 1552–6.

Safir, G. R., Boyer, J. S., and Gerdemann, J. W. (1972). Nutrient status and mycorrhizal enhancement of water transport in soybean. *Pl. Physiol., Lancaster* 49, 700–3.

Saif, S. R. and Khan, A. G. (1977). The effect of vesicular-arbuscular mycorrhizal associations on growth of cereals. III. Effects on barley growth. *Pl. Soil* 47, 17–26.

Sanders, F. E. (1975). The effect of foliar-applied phosphate on the mycorrhizal infections of onion roots. In *Endomycorrhizas* (eds. F. E. Sanders, B. Mosse, and P. B. Tinker) pp. 261–76. Academic Press, London.

Schenck, N. C. and Hinson, K. (1973). Response of nodulating and non-nodulating soybeans to a species of *Endogone* mycorrhiza. *Agron. J.* 65, 849–50.

— and Kellam, M. K. (1978). The influence of vesicular-arbuscular mycorrhizae on disease development. Bull. 798, Agric. exp. Sta. Univ. Florida.

— and Kinloch, R. A. (1976). Mycorrhizal fungi colonizing field crops on a newly cleared woodland site. *Proc. Am. Phytopathol. Soc.* 3, 274.

— and Tucker, D. P. H. (1974). Endomycorrhizal fungi and the development of citrus seedlings in Florida fumigated soils. *J. Am. Soc. Hort. Sci.* 99, 284–7.

Schönbeck, F. (1980). Endomycorrhiza in relation to plant diseases. In *Soil-borne plant pathogens* (eds. B. Schippers and W. Gams). Academic Press, London. (In press.)

— and Dehne, H. W. (1977). Damage to mycorrhizal and non-mycorrhizal cotton seedlings by *Thielaviopsis basicola*. *Pl. Dis. Reptr.* 61, 266–7.

Smith, S. E. and Daft, M. J. (1977). Interactions between growth, phosphate content and N_2-fixation in mycorrhizal and non-mycorrhizal *Medicago sativa* (alfalfa). *Austr. J. Pl. Physiol.* 4, 403–13.

Sparling, G. P. and Tinker, P. B. (1978) Mycorrhizal infection in Pennine grassland. II. Effects of mycorrhizal infection on the growth of some upland grasses on irradiated soils. *J. appl. Ecol.* 15, 951–8.

Stewart, E. L. and Pfleger, F. L. (1977). Development of Poinsettia as influenced by endomycorrhizae, fertilizer and root rot pathogens *Pythium ultimum* and *Rhizoctonia solani*. *Florists Rev.* 159, 37; 79–81.

Sutton, J. C. and Sheppard, B. R. (1976). Aggregation of sand-dune soil by endomycorrhizal fungi. *Can. J. Bot.* 54, 326–33.

Timmer, L. W. and Leyden, R. F. (1978a). Stunting of citrus seedlings in fumigated soils in Texas and its correction by phosphorus fertilization and inoculation with mycorrhizal fungi. *J. Am. Soc. Hort. Sci.* 103, 533–7.

— — (1978b). Relationship of seedbed fertilization and fumigation to infection of sour orange seedlings by mycorrhizal fungi and *Phytophthora parasitica*. *J. Am. Soc. Hort. Sci.* 103, 537–41.

Tinker, P. B. (1980). Role of rhizosphere microorganisms in phosphorus uptake by plants. In *The role of phosphorus in agriculture*. American Society of Agronomy, Madison. (In press.)

Trinick, M. J. (1977). Vesicular-arbuscular infection and soil phosphorus utilization in *Lupinus* spp. *New Phytol.* 78, 297–304.

Vander Zaag, P., Fox, R. L., de la Pena, R. S., and Yost, R. S. (1979). P nutrition of cassava including mycorrhizal effects on P, K, S, Zn and Ca uptake. *Hawaii Agric. Exp. Sta. J.* Series No. 2327.

Warner, A. (1980). Factors affecting the spread of vesicular-arbuscular mycorrhiza. Ph.D. thesis, University of London. (In preparation.)

Wilhelm, S., George, A., and Pendery, W. (1967). Zinc deficiency in cotton induced by chloropicrin-methyl bromide soil fumigation to control *Verticillium* wilt. *Phytopathology* 57, 103 (abstr.).

Witty, J. F. and Hayman, D. S. (1978). Slurry-inoculation of VA mycorrhiza. Rothamsted Report for 1977, Part 1, pp. 239–40.

26 Effects of vesicular-arbuscular mycorrhiza on the size of the labile pool of soil phosphate

E. Owusu-Bennoah and A. Wild

Abstract

The objective was to study the effect of vesicular-arbuscular mycorrhizal infection of the roots of three species (lettuce, onion, and red clover) on the apparent size of the labile pool of soil phosphate in a series of soils. It was hoped that this would help to explain how VA mycorrhizae contribute to the nutrition of plants growing in phosphate deficient soils.

Two soils (Hamble and Sonning) were exhaustively cropped with perennial ryegrass for 30 weeks to reduce the level of labile phosphate. Lettuce, onion, and red clover were grown in these two depleted soils and in undepleted (Batcombe) soil, all of which were labelled with carrier-free isotope ^{32}P. There were four treatments: sterilized soil plus mycorrhiza (S^+), sterilized without mycorrhiza (S^-), non-sterile plus mycorrhiza (N^+), and non-sterile without mycorrhiza (N^-). The plants were harvested 42 days after sowing, the ratios of ^{32}P to ^{31}P taken up by mycorrhizal and non-mycorrhizal plants were measured, and L values calculated.

The mycorrhizal plants took up more phosphorus and became larger than non-mycorrhizal plants in all the soils. Also the mycorrhizal plants grown in Batcombe and Hamble soils had similar specific activities and L values to the non-mycorrhizal plants. It was concluded that mycorrhizal roots do not increase the size of the labile pool in these soils and that both mycorrhizal and non-mycorrhizal roots have access to the same source of phosphorus. In contrast, the mycorrhizal plants (except onion) in Sonning soil had higher L values than their corresponding non-mycorrhizal plants.

A hypothesis is put forward to explain the effects of mycorrhizal infection on the size of the labile pool as measured by the L values. According to this hypothesis there is a slow release of non-labile phosphate in response to high phosphate uptake in the case of mycorrhizal plants, especially when the plant species is one which severely depeletes the soil phosphorus.

27 Species of Endogonaceae and mycorrhizal association of *Elaeis guineensis* and *Theobroma cacao*

P. Nadarajah

Introduction

Spores of the Endogonaceae occur in a wide range of soil types (Mosse 1973). Endomycorrhizae of the vesicular-arbuscular type are widespread (Gerdemann 1968; Mosse 1973) and have been reported in tropical crops such as sugar-cane, coffee, oil palm, cocoa, and tea (Rayner 1939; Laycock 1945; Johnson 1949). In Malaysia, only Wastie (1965) reported the presence of endomycorrhiza and *Endogone* type spores in rubber plantations.

Rayner (1939) reported the presence of vesicular-arbuscular mycorrhiza in oil palm (*Elaeis guineensis* Jacq.) and cocoa (*Theobroma cacao* L.). Pyke (1935) and Laycock (1945) described the endotrophic mycorrhizal association in cocoa roots in Trinidad. A survey of some soil types from oil palm and cocoa plantations was undertaken to study the species of Endogonaceae present.

Materials and methods

Soils were collected from several oil palm and cocoa plantations in the state of Selangor and in one locality in the state of Johore. Each soil sample collected from the uppermost soil layers was mixed thoroughly before examining for the presence of endogonaceous spores. The spores were extracted from the soil using the method of wet-sieving (Gerdemann and Nicolson 1963). Fractions retained on 250, 90, and 45 μm sieves were examined. The size and colour of the spores were estimated in attempting to identify them. The pigmented roots were cleared and stained according to the method of Phillips and Hayman (1970).

Results

Spore types found

Pieces of mycelia were found to be widespread in soil from oil palm and cocoa plantations, but spores were found to be few in some soils.

Most of the spores occurred singly and sometimes in aggregates or sporocarps. The colour of the spores present were white, yellow, orange, reddish-brown, and brown. These were tentatively identified according to spore morphology (Mosse and Bowen 1968) and using the classification of Gerdemann and Trappe (1974). Most of the spores belonged to species of *Glomus*. Parasitism of some spores by other fungi was observed.

Wastie (1965) described dark brown sporocarps obtained from leaf litter under bushes of *Mahogonia macrophylla* grown as ground cover between rubber trees. Similar sporocarps about 300–520 μm in diameter with empty pyriform spores measuring 85–111 × 31–63 μm were observed. Another sporocarp measuring 180–220 μm in diameter, had radially arranged spores around a central gleba. These spores were empty, brown, and clavate, and measured 89–105 × 23–28 μm. The outer wall was less than 1 μm, and at the upper surface the wall was 16–20 μm thick and an inner wall less than 1 μm. At maturity the structure of the inner wall occluded the pore at the point of attachment. These sporocarps probably belong to the genus *Sclerocystis*.

Spores of 120–37 μm diameter, white when young and changing to light brown were found in one site. These spores were attached laterally on a thin-walled hypha with a mother spore at one end (Fig. 27.1). The outer spore wall is perforated (Fig. 27.2), 3–5 μm thick with colourless and membranous inner wall. This belongs to the genus *Acaulospora*.

Spherical spores, yellow to brown, having an average diameter of 118 μm and a wall of 3–6 μm thick were found at another site. Hyphal attachment to this type of spore was not observed. Vesicles borne in clusters were found on the external mycelium (Fig. 27.3) in the soil. These spores and vesicles probably belong to the genus *Gigaspora*.

Most of the soil samples contained spores of the *Glomus* type which were borne mostly free in soil, and sometimes in loose aggregations.

The yellow spores occurring in clusters (Fig. 27.4) were globose with a diameter of 40–55 μm and the wall 3–6 μm thick. Another larger, globose yellow spore, 103–16 μm in diameter, with simple attachment and with an open channel was sometimes recovered from the soil. These spores are similar to *Glomus fasciculatus* (Gerdemann and Trappe 1974).

Different spore types resembling those of *Glomus macrocarpus* were observed. The orange, globose spores (Fig. 27.5) formed singly or in a cluster of 3–5 spores, each 126–45 μm in diameter, might be similar to the variety var. *macrocarpus* (Gerdemann and Trappe

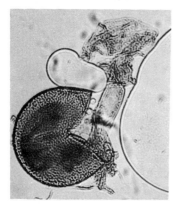

Fig. 27.1. Spore of *Acaulospora* sp attached to empty collapsed vesicle. ×250.

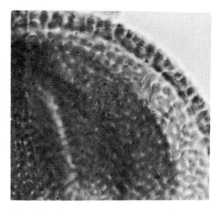

Fig. 27.2. Wall structure of spore in Fig. 27.1. ×950.

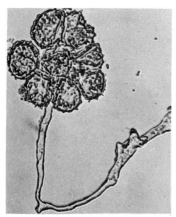

Fig. 27.3. Vesicle of *Gigaspora* sp. × 214.

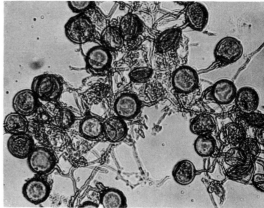

Fig. 27.4. Aggregates of yellow vacuolate spores, probably *Glomus fasciculatus.* × 150.

1974). A wide-necked spore occurring singly in soil, globose to ovoid; wall 4-6 μm thick, sometimes with radial pores was similar to that described by Mosse and Bowen (1968). Gerdemann and Trappe (1974) have included it in the above variety. A reddish-brown spore probably belonging to the same variety formed singly in soil. The spore is globose, 260-319 μm diameter. The outer wall is hyaline, 2-4 μm thick and inner coloured wall 7-10 μm thick. The ornamented wall is light to reddish brown.

In one site reddish brown to dark brown spores were found singly in the soil. These spores were globose to ellipsoid measuring 70-107

X 79-120 μm, outer wall hyaline, less than 1 μm and inner wall being 7-10 μm thick. These spores correspond to *G. macrocarpus* var. *geosporus* as described by Gerdemann and Trappe (1974) and Hall (1977).

Description of the mycorrhiza

The mycorrhizal and non-mycorrhizal roots could not be differentiated from their outward appearances as they were heavily pigmented. The tertiary and quarternary roots of cocoa (Laycock 1945) and oil palm were observed to be mycorrhizal.

The external mycelium is dimorphic with thick-walled coenocytic hyphae and thin-walled septate hyphae and is in accordance with that described by Butler (1939), Mosse (1959), and Wastie (1965) for other plants. Frequent development of septation in certain thick-walled external mycelium was noted (Mosse 1959). On some of the external hyphae, terminal vesicles were found (Figs. 27.6 and 27.7). They were globose to ovate with dense contents when young. The walls of the vesicles were slightly thicker than the walls of the bearer hyphae.

Penetration of roots by thick-walled hyphae occurs through the epidermis. At the point of entry, the hypha is slightly swollen forming an appressorium. Then the hypha develops longitudinally, the cortex is colonized by hyphae forming fungal coils almost filling the cell lumina. Arbuscules were not seen in both roots but Wastie (1965) noted some arbuscules in rubber roots. As reported by Laycock (1945) for cocoa and Wastie (1965) for rubber, the arbuscules present were accompanied or replaced by fungal coils. Vesicles were common in cortical cells (Fig. 27.8). The terminal vesicles were either inter- or intracellular and spherical to ellipsoidal. There is a free connection between the cytoplasm of the parent hyphae and the vesicle. The wall of the vesicle is the same thickness as the hypha.

Discussion

The survey of different soils confirms that endogonaceous spores occur widely in soils around oil palm and cocoa. Most of the spores extracted were from the top 15 cm of the soil as also observed by Redhead (1977). Wastie (1965) reported four types of spores in the soils growing around rubber roots. In soils growing around oil palm and cocoa roots, most of the spores recovered were *Glomus* type and genera of two types of *Sclerocystis* and one each of *Acaulospora* and *Gigaspora*. In some sites up to four different spore types occurred together and in others these were dominated by one spore type.

The fungus examined did not infect the vascular tissues of the

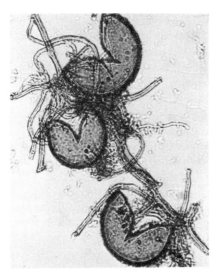

Fig. 27.5. Orange spores, probably of *G. macrocarpus* type. × 225.

Fig. 27.6. External vesicle of *Elaeis guineensis.* × 250.

Fig. 27.7. Mycorrhiza of *Theobroma cacao*, external vesicle. × 214.

Fig. 27.8. Vesicles of *E. guineensis* in cortical cells. × 214.

roots but occurred in the cortex only. Stages of coils of the fungi and vesicles were observed in the roots and it is possible that more than one species of *Glomus* is responsible for the mycorrhizal infection. Pyke (1935) applied the term 'vesicle' to the appressorium formed on entry of the external hyphae. Laycock (1945) did not see any vesicles in the cocoa roots in Trinidad. In the present study, the vesicles in oil palm and cocoa roots were observed. The spores of the other genera were not a regular component and were not associated with the roots at the time of observation.

Pot cultures (Gerdemann 1955) are being tried and identification will be confirmed as they mature.

References

Butler, E. J. (1939). The occurrence and systematic position of the vesicular-arbuscular type of mycorrhizal fungi. *Trans. Br. mycol. Soc.* 22, 274–301.

Gerdemann, J. W. (1955). Relation of a large soil-borne spore to mycorrhizal infections. *Mycologia* 47, 619–32.

— (1968). Vesicular-arbuscular mycorrhiza and plant growth. *A. Rev. Phytopathol.* 6, 397–418.

— and Nicolson, T. H. (1963). Spores of mycorrhizal *Endogone* species extracted from soil by wet-sieving and decanting. *Trans. Br. mycol. Soc.* 46, 235–44.

— and Trappe, J. M. (1974). The Endogonaceae in the Pacific Northwest. *Mycol. Mem.* 5, 1–76.

Hall, I. R. (1977). Species and mycorrhizal infections of New Zealand Endogonaceae. *Trans. Br. mycol. Soc.* 68, 341–66.

Johnson, A. (1949). Vesicular-arbuscular mycorrhiza in sea island cotton and other tropical crops. *Trop. Agric., Trin.* 26, 118–21.

Laycock, D. H. (1945). Preliminary investigations into the function of the endotrophic mycorrhiza of *Theobroma cacao* L. *Trop. Agric., Trin.* 22, 77–80.

Mosse, B. (1959). Observations on the extra-matrical mycelium of a vesicular-arbuscular endophyte. *Trans. Br. mycol. Soc.* 42, 439–48.

— (1973). Advances in the study of vesicular-arbuscular mycorrhiza. *A. Rev. Phytopathol.* 11, 171–98.

— and Bowen, G. D. (1968). A key to the recognition of some *Endogone* spore types. *Trans. Br. mycol. Soc.* 51, 469–83.

Phillips, J. M. and Hayman, D. S. (1970). Improved procedures for clearing roots and staining parasitic and vesicular-arbuscular mycorrhizal fungi for the rapid assessment of infection. *Trans. Br. mycol. Soc.* 55, 158–61.

Pyke, E. E. (1935). Mycorrhiza in cacao. Rep. Cacao Res., Trin. (1934), pp. 41–8.

Rayner, M. C. (1939). The mycorrhizal habit in crop plants, with a reference to cotton. *Emp. Cott. Grow. Rev.* 16, 171–9.

Redhead, J. F. (1977). Endotrophic mycorrhizas in Nigeria: species of the Endogonaceae and their distribution. *Trans. Br. mycol. Soc.* 69, 275–80.

Wastie, R. L. (1965). The occurrence of an *Endogone* type of endotrophic mycorrhiza in *Hevea brasiliensis*. *Trans. Br. mycol. Soc.* 48, 167–78.

28 Mycorrhizae of *Hevea* and leguminous ground covers in rubber plantations

U. P. de S. Waidyanatha

Description of the mycorrhizae

Hevea brasiliensis and legumes such as *Pueraria phaseoloides, Centrosema pubescens, Calopogonium mucunoides, Desmodium ovalifolium*, and *Stylosanthes guianensis* grown as ground cover under rubber are infected with VA mycorrhizae. Wastie (1965) has studied in some detail the mycorrhizal association in *Hevea* and showed that the endophytes are of the *Endogone* type, essentially similar to that described previously by other workers. Preliminary observations on the infection and morphological features of the endophytes with *Hevea* are generally consistent with those of Wastie (1965).

Infection appeared to be more extensive in surface-feeding rubber roots from juvenile (1–5-year-old trees) plantations with good undergrowth of creeping legumes, mainly *Pueraria*. The rootlets here form a network within the first few inches of the soil and the decaying leaf debris from the legume creepers. Secondary laterals invading deeper soil layers in such situations as well as surface feeding rootlets in mature plantations (with little or no undergrowth) appear to be relatively less infected. The soil rich in organic carbon and nutrients beneath legume covers and the higher humidity in that situation seem to provide an ideal environment for spread of the mycorrhizal fungi and for invasion of host roots.

Our observations of the mycelial morphology and behaviour of the fungi inside the rubber roots are again similar to those of Wastie (1965), except perhaps that arbuscules appear to be very infrequent. The morphological features of the endophytes in cover legumes appear visually similar to that of *Hevea*, suggesting that the same endophytes indiscriminately infect both types of host. Unsuccessful attempts were made to trace hyphal connections between rubber and cover legume roots. Such a finding, while establishing that the same endophyte infects both hosts, raises the interesting possibility of interspecific translocation of nutrients through hyphae between the hosts.

Wastie (1965) reported sporocarps and four types of resting spores from rubber growing soils of Malaysia. Our observations on sporocarps and spores have been confined so far to few soil samples from each of four rubber growing localities. However, a wide range of about 18 spore types were observed. Many of them are essentially similar to those that have been described by other workers. Many of the spore types have been assigned to genera on the basis of the classification of Gerdemann and Trappe (1974), but some still remain to be identified at species level. *Glomus* was the most predominant genus at all four sites. Out of about eight *Glomus* types, *G. fasciculatus, G. mosseae,* and *G. microcarpus* were identified. *Sclerocystis* species were the next most common; there being about five types. Two types of *Acaulospora* and two of *Gigaspora* were also observed. These observations confirm the world-wide distribution of these genera of the family *Endogonaceae*.

Of the four locations, three were in the Wet Lowlands of Sri Lanka receiving a rainfall of 3000–3800 mm annually with no marked dry spells, the fourth was in the Dry Zone with a rainfall of 1800–2500 mm and a prolonged drought. Spore numbers were much higher in the latter location and was of the order of 2000–2800 per 100 g air-dried soil. At the other three sites, spore numbers varied from about 400 to 1100. It is probable that the prolonged dry period in the Dry Zone favours spore formation, whereas humid conditions prevailing for most of the year in the Wet Zone favour vegetative growth of the endophytes. The spore numbers are among the highest recorded in the literature and confirm the abundance of VA mycorrhizae in the humid tropics (Redhead 1977).

The effect of mycorrhizae on growth, phosphorus uptake, and nitrogen fixation of cover legume

Pueraria, Centrosema, Calopogonium, Desmodium, and *Stylosanthes* failed to establish or were very stunted when grown in sterilized soil. This was ascribed to the absence of mycorrhizal propagules because inoculation of the soil with surface-sterilized mycorrhizal spores restored normal growth. It therefore appears that, under natural conditions, mycorrhizal infection is an obligatory condition for growth of the leguminous covers.

Further work (Waidyanatha, Yogaratnam, and Ariyaratne 1979) revealed that growth of *Pueraria* in a methylbromide-treated lateritic clay loam (pH 4.5), deficient in available phosphorus (3.8 p.p.m. NH_4F/HCl extractable phosphorus), was very poor unless the plants were inoculated with mycorrhizae or supplied with 500 mg of rock phosphate (RP) per kg of soil. 100 mg RP/kg soil did not improve growth above that of the untreated control. The concentration of phosphorus in mycorrhizal plants was comparable with that on non-mycorrhizal plants receiving rock phosphate at the higher level, clearly showing that the presence of the endophyte greatly enhanced phosphorus extraction from the soil. The growth of inoculated plants was only slightly better than the phosphate-fertilized plants, whereas nodule weights, numbers, and acetylene reduction activity were much greater in the former than in the latter. This suggests that mycorrhizal infection stimulated nitrogen fixing activity relatively more than plant growth. It is difficult to explain how mycorrhizae stimulate the nitrogen fixing system other than through better phosphorus nutrition of the plant, but the data emphasize the need for deeper scrutiny of this double symbiosis in legumes.

Mycorrhizae on growth of *Hevea*

Some investigations were made to ascertain whether mycorrhizal infection is a pre-requisite for growth of *Hevea* by growing seedlings in pots of methylbromide-fumigated or untreated (normal) soil deficient in phosphorus. Methylbromide had not, in this instance, killed all mycorrhizal propagules as shown by infections observed in some plants grown in the fumigated soil, although the amount of infection was far less than in the non-fumigated soil. There was no difference in growth between the two sets of plants. Even a comparison of growth in completely uninfected pots with that in highly infected pots did not reveal any clear differences. It appears, therefore, that mycorrhiza is not an absolute requirement for growth of *Hevea*, as is the case with the legume crops.

Proposed research

The following research objectives are envisaged and work is in progress with the view to ascertain whether mycorrhizae of *Hevea* and legume covers could be manipulated for greater crop productivity:
1. A survey of the endophyte types in local soils.
2. Evaluation of these types and others in sterilized soil on the basis of growth responses on the host and nutrient uptake for possible selection of better ones.
3. Performance of the selections in normal soil in competition with native types in the soil.
4. Economic feasibility of inoculation on a field scale with promising types.

References

Gerdemann, J. W. and Trappe, J. M. (1974). The Endogonaceae in the Pacific Northwest. *Mycol. Mem.* 5, 1–76.
Redhead, J. F. (1977). Endotrophic mycorrhizas in Nigeria: species of the Endogonaceae and their distribution. *Trans. Br. mycol. Soc.* 69, 275–88.

Waidyanatha, U. P. de S., Yogaratnam, N., and Ariyaratne, W. A. (1979). Mycorrhizal infection on growth and nitrogen fixation of *Pueraria* and *Stylosanthes* and uptake of phosphorus from two rock phosphates. *New Phytol.* (In press).

Wastie, R. L. (1965). The occurrence of an *Endogone* type of endotrophic mycorrhiza in *Hevea braziliensis*. *Trans. Br. mycol. Soc.* 48, 167–78.

29 Prospects of mycorrhizal inoculation of tin-tailings of Malaysia

M. N. Shamsuddin

Introduction

Most of the wastelands in Malaysia result from tin mining activities. They occur mostly along the western range, but less extensive fields occur along the eastern edges of the central mountains. The total mining area is approximately 146 000 ha and half of this area is now a waste environment of tin-tailings (Palaniappan 1974). Many towns in West Malaysia grew up as a result of tin mining activities, and owing to rapid urban expansion many of the disused mining lands are now found on the fringe of cities. Their close proximity to densely populated areas makes them valuable for exploitation for food production, reforestation for recreational purposes, and for housing.

Owing to the poor conditions of the wastelands, the need for rehabilitation is obvious. The first attempt to revegetate the area was carried out by Birkenshaw (1931) and followed later by Mitchell (1957). Graham, Tan, and Vythilingam (1975) and Maene, Mok, and Ling (1977) investigated the possibility of using the area for growing fruit and vegetables. For successful plant growth, heavy fertilization and addition of large amounts of organic matter are required. The importance of rhizobium and mycorrhiza inoculation of the waste area has not previously been investigated.

Soil properties of tin-tailing

These are generally sandy areas with a thorough drainage system by which most of the soluble minerals are lost. Such tailings have very low mineral content and cation exchange capacity (CEC) ranges from 0.4 to 3.1. This is considered to be one of the most serious limitations for agricultural development.

The slime retention areas have too high a proportion of clay particles, and as a result, lack porosity. Such areas are characterized by dried clay masses and water-logged conditions with a relatively higher organic and CEC value. Negligible amounts of clay mineral and organic matter in the sandy areas result in a certain degree of chemical and biotic inertness. Such conditions can be described as unique in a humid tropical environment which usually favours microbial activities including a wide spectrum of plant pathogenic organisms.

This inherent biotic 'sterility'of sand tin-tailings may therefore be exploited for the cultivation of disease susceptible crops. It could also offer less competition to desirable micro-organisms such as rhizobium and mycorrhizal fungi when they are introduced into the inert environment.

Natural occurrence of mycorrhiza in tin-tailings

Plant regeneration in mined-over areas depends to a large extent on the structure and composition of the tailings and the mining methods employed. Available evidence indicates that slime retention dams are often colonized by a variety of plant species and that regeneration may be quite rapid in these areas (Reid 1956; Mitchell 1964; Palaniappan 1974). After a period of 30 years the species composition of slime-tailings forest is similar to that of secondary forest (Palaniappan 1972, 1974).

A survey of mycorrhizal infection of roots of the pioneer plants on tailings was conducted (Shamsuddin, unpublished). It was found that tailings of about five years old were free from mycorrhiza and rhizobia. On the other hand, plants from well colonized tailings were found to be infected. It is possible that reinfection of sterile tailings could come from adjacent areas through water and wind-blown soil.

Mycorrhizal inoculation of tailings

The organic matter added to the tailings is usually in the form of peat which is available locally. Waste products such as rice straw, oil-palm pericarp waste, cow dung, chicken dung, and even human excreta are used for intensive cultivation of vegetables and fruit trees.

The animal waste such as chicken dung is free from mycorrhizal fungi. In an experiment on tin-tailing where three types of organic matter were added, it was found that plants grown on peat and sewage sludge were mycorrhizal, whereas plants grown on tailing only, or with chicken dung added, were non-mycorrhizal (Shamsuddin, unpublished). It was also observed that transplanted young seedlings grown on the 'sterile' tailing were mycorrhizal.

From this initial observation it can be said that there is a possibility of successful mycorrhizal inoculation of plants on tin-tailing. Also, because of the sterile nature of the tailing, selective mycorrhizal fungi could be introduced successfully without interference with the endogenous mycorrhizal population. Of course the physical and chemical nature of the tailings will determine not only the establishment of the fungi but also the response of plant infection under these conditions. Immediate application of mycorrhizal inoculation of seedlings or plants to be planted on tin-tailing could be carried out under certain intensive practice such as in the growing of fruit trees.

References

Birkenshaw, F. (1931). Reclaiming old mining land for agriculture. *Malay agric. J.* 19, 470–6.

Graham, K. M., Tan, H., and Vythilingam, S. (1975). Cultivation of vegetables on tin-tailings and supplemented by oil palm pericarp waste. *Malay agric. Res.* 4, 155–8.

Maene, L. M., Mok, C. K., and Ling, A. H. (1977). Reclamation studies on Malaysian tin tailings for crop production. Proc. Conf. on Malaysian Food Self-sufficiency, pp. 197–206.

Mitchell, B. A. (1957). Malayan tin-tailings—prospects of rehabilitation. *Malay. Forester* 20, 181–6.

—— (1964). The ecology of tin mine spoil heaps. *Malay. Forester* 22, 111–32.

Palaniappan, V. M. (1972). History, flora and edaphology of tin-tailing area around Kuala Lumpur. *Trop. Ecol.* 13, 202–25.

—— (1974). Ecology of tin tailings areas: plant communities and their succession. *J. appl. Ecol.* 11, 133–50.

Reid, J. A. (1956). Plants of tin-tailings. *Malay. Nat. J.* 12, 3.

30 Influence of 1, 2-dibromo-3-chloro-propane (DBCP) fumigation on mycorrhizal infection of field-grown groundnut

G. Germani, H. G. Diem, and Y. R. Dommergues

Abstract

An experimental design was established at Patar, Senegal in a typical sandy soil. It consisted of nine randomized blocks, with two treatments: fumigated with DBCP 14 days before planting and non-fumigated. Treated plots were fumigated with DBCP at a depth of 20 cm 14 days before sowing at a rate of 25 l/ha. Groundnut cultivars 28-206 (growth-cycle 120 days), 55-437 (growth cycle 90 days), and GH 119-20 (growth cycle 120 days) were planted on 23 July 1977. During the groundnut growth cycle, rainfall was low (287 mm during June–October), but well distributed. Root systems and soil were sampled twice (16th and 60th day) in order to estimate endomycorrhizal infection, N_2 (i.e. C_2H_2) fixation, and nematode numbers. The crop was harvested on 20 October 1977 (cv. 55-437) and 10 November 1977 (cvs. 28-206 and GH 119-20).

DBCP fumigation totally suppressed nematode populations consisting mostly of Scutellonema cavenessi. Mycorrhizal infection, as measured at the first sampling (16th day), was significantly higher in fumigated plots for one cultivar (28-206). Since the roots of groundnuts cvs. 28-206 and GH 119-20 were heavily infected in the fumigated plots as early as the 16th day, endomycorrhizal infection of some cultivars appeared to be a precocious process in the edaphic and climatic conditions of Senegal, provided that there is no limiting factor, such as nematode injury. No significant difference between treated and untreated plots was found at late sampling (60th day). Microscopic observations showed that only very few vesicles and arbuscules were found in roots of groundnut, as reported by Ross and Harper (1973).

DBCP fumigation improved N_2 fixation of cultivars 55-437 and GH 119-20 but not that of cv. 28-206. Whereas endomycorrhizal infection of cv. 28-206 was not improved by fumigation, its N_2 fixation was greatly enhanced by this treatment, at least during the period ranging from day 40 to 50 (Germani 1979).

DBCP fumigation increased phosphorus content of groundnut pods by 134, 99, and 146 per cent and nitrogen content (pods) 127,

Table 30.1. Effect of DBCP fumigation on mycorrhizal infection of field-grown groundnut expressed as percentage of infected roots

	Groundnut cultivar		GH 119–20
	55–437	28–206	
First sampling (16th day)			
No fumigation	28.4 a (a)	22.8 a (a)	53.8 a (b)
Fumigation	33.0 ab (a)	59.6 b (a)	76.5 a (b)
Second sampling (60th day)			
No fumigation	54.0 ac (a)	63.6 b (a)	57.0 a (a)
Fumigation	56.2 c (a)	40.7 ab (a)	57.4 a (a)

Numbers in columns not having same letter and numbers in rows not having same letter between brackets differ *P* = 0.05 by Mann–Whitney test.

135, and 133 per cent respectively for cultivars 55–437, 28–206, and GH 119–20.

These results support the conclusion of Bird, Rich, and Glover (1974) that nematodes might eventually limit mycorrhizal infection, but the related mechanism is still unknown. Nematodes could also impede the establishment and functioning of the double plant symbiosis (Mosse, Powell, and Hayman 1976; Daft and El-Giahmi 1976) (a) with endomycorrhizae (b) with *Rhizobium*, thus indirectly reducing plant phosphorus uptake, and possibly its water uptake (Safir, Boyer, and Gerdemann 1977), and also its nitrogen-fixing ability. Such consequences are presumably most harmful to the plant, especially in phosphorus and nitrogen deficient soils, which commonly occur in semi-arid West Africa. The hypothesis that DBCP could directly enhance endomycorrhizal infection, as suggested by Menge (personal communication) should also be taken into account.

References

Bird, G. W., Rich, J. R., and Glover, S. U. (1974). Increased endomycorrhizae of cotton roots in soil treated with nematicides. *Phytopathology* 64, 48–51.

Daft, M. J. and El-Giahmi, A. A. (1976). Studies on nodulated and mycorrhizal peanuts. *Ann. appl. Biol.* 83, 273–6.

Germani, G. (1979). Action directe et résiduelle d'un traitement nématicide du sol sur trois cultivars d'arachide au Sénégal. *Oléagineux* (submitted).

Mosse, B., Powell, C. L., and Hayman, D. S. (1976). Plant growth responses to vesicular-arbuscular mycorrhiza. IX. Interactions between VA mycorrhiza, rock phosphate and symbiotic nitrogen fixation. *New Phytol.* 76, 331–42.

Ross, J. P. and Harper, J. A. (1973). Host of a vesicular-arbuscular *Endogone* species. *J. Elisha Mitchell scient. Soc.* 89, 1–3.

Safir, G., Boyer, J. S., and Gerdemann, J. W. (1971). Mycorrhizal enhancement of water transport in soybean. *Science, N.Y.* 172, 581–3.

31 Selection of mycorrhizal isolates for biological control of *Fusarium solani* f. sp *phaseoli* on *Vigna unguiculata*

S. N. Baradas and P. M. Halos

Abstract

Isolates of vesicular-arbuscular mycorrhizae belonging mainly to *Glomus mosseae, G. macrocarpus* var. *macrocarpus, G. macrocarpus* var. *geosporus, G. fasciculatus*, and few to *Gigaspora* spp were assayed for stimulatory influence on growth and protection against *Fusarium* dry root rot of cowpea (*Vigna unguiculata*). These isolates were collected from different plant roots along with rhizosphere soils, namely acacia (*Acacia* sp), adzuki bean (*Phaseolus pubescens*), avocado (*Persea americana*), bean (*Phaseolus vulgaris*), bamboo (*Bambusa blumeana*), banana (*Musa sapientum*), bottle gourd (*Lagenaria siceratia*), chico (*Achras sapota*), citrus (*Citrus sinensis*), cacao (*Theobroma cacao*), cassava (*Manihot utillisima*), coconut (*Cocos nucifera*), coffee (*Coffea robusta*), corn (*Zea mays*), cotton (*Gossypium hirsutum*), cowpea (*Vigna unguiculata*), egg-plant (*Solanum melongena*), ginger (*Zingiber officinale*), grape (*Vitis vinifera*), guinea grass (*Panicum maximum*), okra (*Abelmoschus esculentus*), batao (*Dolichos lablab*), ipil-ipil (*Leucaena glauca*), jackfruit (*Artocarpus heterophyllus*), kamias (*Averrhoa bilimbi*), lanzones (*Lansium domesticum*), palm tree (*Roystonea elata*), mung bean (*Phaseolus radiatus*), mango (*Mangifera indica*), lima bean (*Phaseolus lunatus*), papaya (*Carica papaya*), peanut (*Arachis hypogea*), santol (*Sandoricum koetjape*), pigeon pea (*Cajanus cajan*), rice (*Oryza sativa*), soyabean (*Glycine max*), squash (*Cucurbita maxima*), caimito (*Chrysophyllum cainito*), star fruit (*Averrhoa carambola*), sugarcane (*Saccharum officinarum*), sweet pepper (*Capsicum annum*), sweet potato (*Ipomea batatas*), tamarind (*Tamarindus indica*), gabi (*Colocasia esculenta*), tobacco (*Nicotiana tabacum*), tomato (*Lycopersicon lycopersicum*), water sponge (*Luffa aegyptiaca*), and winged bean (*Psophocarpus tetragonolobus*).

Heat sterilized and non-sterilized soil media were used in growing cowpea. For sterilized soil, two pots with three plants each, were inoculated by mixing seeds and the surrounding soil thoroughly with extracted mycorrhizal vesicles and infected roots while the other two pots, also with the same number and kind of seeds, were without mycorrhiza. The inoculated extra-matrical vesicles of *Glomus* and *Gigaspora* spp were taken from rhizosphere soils by wet-sieving and decanting. Plants in the pots were inoculated five days later with pure cultures of *Fusarium solani* f. sp *phaseoli*, a pathogen causing dry root-rot on cowpea. Control treatments consisted of similar plants without mycorrhizal or *Fusarium* inoculation, or either mycorrhizal or *Fusarium* inoculation alone. The same set of treatments were maintained for non-sterilized soil.

Data analysed included the effect of the different isolates on the initial and final heights of plants in both sterilized and non-sterilized soil media and the effect of strain and soil interaction on the incidence of *Fusarium* dry root-rot infection. The effect of the individual mycorrhizal inoculations on growth rates and incidence of *Fusarium* dry root-rot infection were analysed using the two-factor factorial in CRD analysis of variance. Some promising mycorrhizal isolates were selected although the significance is doubtful as factors like the soil and strain interaction must be considered. For instance, certain strains may be efficient in sterilized soil where there are not many competing micro-organisms but behave differently in non-sterilized soil where they interact with other soil microflora. Nevertheless, a close analysis of the results using Duncan's multiple range test revealed eleven promising mycorrhizal isolates, namely, those derived from tomato, sweet pepper, rice, soyabean, pigeon pea, coffee, ginger, guinea grass, jackfruit, lima bean, and mung bean. To a certain extent, mycorrhizal fungi increased the final height of inoculated plants, probably due to an enhanced ability of infected roots to absorb water and nutrients, since mycorrhiza can act as an auxiliary absorbing system. Generally, growth response of cowpea plants is much less in unsterilized than in sterilized soils. Successful *Fusarium* infection resulted in stunting, chlorotic leaves, brownish stem bases, and shrivelled roots of cowpea. Prior inoculation of seeds with isolates of *Glomus* and *Gigaspora* spp from lima bean, soyabean, and ginger gave the least mean percentage of pathogen infection indicating a greater protection of cowpea against the disease. Probably these mycorrhizal isolates and the associated microflora acted as competitors or provided mechanical and chemical barriers against the establishment of *Fusarium solani* f. sp *phaseoli* on cowpea.

PART VI

Concluding remarks

Concluding remarks

The material of the preceding pages was presented at the International Workshop on Tropical Mycorrhiza Research in Kumasi, Ghana, in August–September 1978. This workshop reported the recent progress in the knowledge on mycorrhizae in the tropics and also clearly revealed the urgent need for increased research in the area.

Different aspects of mycorrhiza research are dealt with in chronological order. Thus, as the importance of mycorrhizal symbiosis was first experienced at the introduction of pines into the tropics, papers on pine afforestation and ectomycorrhizal inoculation in particular constitute the first section. Lots of experience has been gained in this area and also applied to forestry practice but, considering the growing importance of intensive forestry for the development of many tropical countries, the need for further research is quite evident. Dr D. H. Marx, the Chairman of this session at the workshop, crystallized the present knowledge and future needs in following five points:

(i) There is little doubt that ectomycorrhizal fungi contained in variously-obtained pine soil inocula are responsible for the successful establishment in the nursery; ectomycorrhizal development is consistently found on roots of the seedlings.

(ii) Research efforts should continue to ascertain the identity of the fungi forming ectomycorrhizae in soil inoculum. In certain cases, one morphological type is consistently observed. In others, it appears that more than one fungus species may be present. In order for us to understand the role of these fungi we must have their identity so we can relate to previous work and communicate to each other with greater accuracy.

(iii) It would be ecologically more sound if a mixture of ectomycorrhizal fungi were present on seedlings. The ultimate objective should be to have as many good fungi forming ectomycorrhizae on trees in the field as possible. This should ensure that as environmental conditions change during the year (e.g. temperature and rainfall) the trees will consistently have a physiologically active root system essential for them to maintain acceptable vigour.

(iv) Numerous opportunities exist for research to be accomplished on pure culture introduction of selected fungi. Not only would this type of research furnish vital information as to which of the test fungi are good candidates for a practical and operational programme of afforestation, but also furnish information on which of the fungi are not effective.

It is just as important to know where a fungus is *not* effective as to know where it is effective.

(v)   Research results are only as good as the techniques used to obtain them. A variety of techniques have been discussed during the workshop that work well for intended objectives. Even though laboratory facilities and materials vary from country to country, the soundness of the techniques should not be altered. In nursery and field tests, non-inoculated plants must be used as a treatment in order to assess the possible presence of naturally occurring fungal symbionts. It also helps to ascertain the growth significance due to inoculation. Another major need is to assess the amount of mycorrhizae formed on seedlings. In certain instances the amount of ectomycorrhizae on roots at the time of planting may be as significant as the species of fungus forming the mycorrhizae.

In comparison with exotic pines, rather little is known on mycorrhizal relations in the natural vegetation of the tropics. This open field of research was summarized at the closing session of the workshop by Dr J. F. Redhead as follows:

Recent work, including the papers presented at this workshop, confirms previous studies that the vesicular-arbuscular mycorrhizal association is by far the commonest type found in the tropical rain forests. The paper by de Alwis and Abeynayake is of particular interest, as it confirms studies by Singh that ectomycorrhizal association occurs widely in the important Asian family Dipterocarpaceae.

Further surveys are needed in important tropical vegetation types such as the 'Miombo' woodland which stretches from Tanzania in East Africa to the Atlantic Ocean in Angola. Further studies should be made of families such as the Euphorbiaceae and Rhamnaceae in which ectomycorrhizal association have been reported and also to investigate whether savannah species of Caesalpiniaceae form this type of association.

The workshop has focused attention on the need to identify and classify important mycorrhizal fungi occurring naturally in the tropics. These fungi can then be tested in field trials on important crop plants in agriculture and forestry.

Particular interest was shown in the occurrence of natural ectomycorrhizal associations. The fungi involved are obviously suited to tropical conditions of high temperatures and, in places, long, severe dry seasons. Saleh-Rastin and Djavanshir reported the occurrence of several ectomycorrhizal fungi in the natural forests of Iran. As this country is subject to extremes of temperature, low rainfall and a long dry season these fungi are of special interest as possible mycorrhizal associates for *Pinus* species now being widely planted in tropical countries. Attempts are now in progress to identify such mycorrhizal fungi which may be used in tropical afforestation. These fungi may be much more efficient symbionts than varieties of temperate species which are often not well suited to a lowland tropical environment.

Technical problems in studying mycorrhizae in tropical forests were emphasized in the paper by Riess and Rambelli. Great difficulty is found in tracing mycorrhizal roots back to the parent tree. Another difficulty is experienced in isolating mycorrhizal fungi from the roots. This process is necessary in relating the mycorrhizal fungus to associated sporophores and for re-synthesis of the association. The need for adequate controls in all pot and field experiments was also emphasized.

From these morphological and taxonomic aspects of mycorrhizae in the tropics the discussion continues to their physiology, particularly the role of the association for the nutrition of tropical plants. The present knowledge and prospects were reviewed by Dr G. D. Bowen as follows:

There are now an increasing number of records of beneficial responses of tropical plants to inoculation with ecto- and endomycorrhizal fungi. Ecto- and VA fungi are functioning in a remarkably similar manner in spite of great diversity of structure and the fungi involved.

Generally reports focus on improved nutrient uptake which occurs mainly via fungal growth into soil, the absorption of nutrients and transfer to the plant. Most studies refer particularly to phosphate nutrition but it should be remembered that increased absorption of other ions, e.g. zinc is also important in many soils. We should not focus on phosphate alone.

Apart from the fertility level of the soil being studied major factors in response of a plant species to mycorrhizal infection are rooting intensity of the species and the growth of the fungus into the soil. Little is known of how soil and management factors affect fungal growth and it is an area deserving much more study in order to optimize the mycorrhizal responses. Data also are lacking on the relation of the development of the infection to the development of the root system and nutrient demand.

Fungus growth in and around the roots raises the question of the assimilate cost to the plant. In general the benefits from associations far outweigh the assimilate cost and anyway the plant may well compensate for any extra assimilate going to the roots, although this latter point has not been studied. However, in some fertility situations where the mycorrhizal infection is still high depression of growth has been recorded. Such fertility situations are not likely to be common in tropical soils but we should be alert to the possibility that high fertilizing may sometimes be counterproductive.

It is now clear from several studies, most of which are on VA mycorrhizae, that the mycorrhizae do not 'mine' sources of phosphate unavailable to the non-mycorrhizal plants, but that they absorb available phosphate more efficiently. Thus, sustained mycorrhizal responses cannot be expected without some phosphate input, probably lower than with non-mycorrhizal plants.

The main economic interest in the mycorrhizal response is in increased absorption of nutrients which are in short supply in the soil. However, research is needed into other possible roles of fungal growth into soil, especially where conditions reduce root growth. The fungus may give the root an alternate strategy for many functions, particularly in stress situations. Mycorrhizal function in ameliorating the effects of decreased rooting due to high salinity, high aluminium and other harmful elements, and root disease deserve more attention.

Maximizing the efficiency of the use of soil nutrients involves the *use* of nutrients in the plant as well as *uptake*, the focus of most studies. There is increasing evidence for both ecto- and endomycorrhizae that differences other than nutritional ones between mycorrhizal and non-mycorrhizal plants sometimes occur. One example of this was given by Moawad—mycorrhizal plants had quite different growth–temperature relationships from non-mycorrhizal plants. This may be of special importance in the tropics where high soil temperatures are often experienced. Similarly the work of Marx suggests that the ectomycorrhizal fungus *Pisolithus tinctorius* may be particularly useful at high soil temperatures.

Finally, studies on mycorrhizal function have been directed towards single species. Evidence for a major role of mycorrhizal fungi in *community dynamics* is now emerging, e.g. in assisting tightly closed nutrient cycles in perennial plant systems and in competition between species in man-made and natural mixed ecosystems. The mycorrhizal factors may be important in optimizing yield in multiple and mixed cropping of the tropics. Mixed systems also introduce other considerations such as allelopathy and effects of mutual shading, an exciting field for further study.

As a whole, mycorrhizal systems are so common and their effects so marked that many of the traditional nutritional and physiological concepts of how plants work will have to be reconsidered.

From this speculation we come automatically to the question how to apply the present knowledge on mycorrhizal symbiosis to tropical agriculture and what are the prospects for its better utilitzation. This was the topic of the last session of the workshop. When summarizing the discussion Dr Barbara Mosse pointed out, among other things, following essential aspects:

Agricultural plants comprise annuals, herbaceous plants, and perennials, and nutritional requirements of such plants differ not only between these groups but also between species within each group. Furthermore, nutrient requirements vary at different stages of plant development. By understanding the different requirements and the particular problems associated with each crop it might be possible to influence the mycorrhizal contribution to plant growth, remembering that essentially mycorrhiza function within a three-component system consisting of plant, fungus, and soil. Some of these interrelationships were illustrated and results of some small-scale pilot experiments conducted under field conditions with maize, wheat, potato, citrus, soyabean, and clover were discussed.

Of special interest in VA mycorrhiza research is the case of legumes and the effect of the fungus on nodulation and symbiotic nitrogen fixation by rhizobia. By providing the phosphorus essential for symbiotic nitrogen fixation VA mycorrhiza may improve the competitive ability of legumes in pastures and may improve the establishment and value of a leguminous cover crop such as occurs in rubber plantations. There is some circumstantial evidence that mycorrhizal and nitrogen fixing symbiotic systems of legumes may interact synergistically in other ways than through phosphate nutrition and may affect the protein content of the crop.

Finally various pitfalls in techniques of conducting experiments were discussed.

For summing up the main achievements of the workshop, following quotation from the speech of Mr M. A. Chaudhry of Uganda, speaking at the concluding session on behalf of the participants, may be adequate:

(i) The young scientists have been enabled to build up clear concepts and to have proper orientation.
(ii) We developed a sense of belonging to an international brotherhood. It was only mental before, we have now got a physical touch of it.

(iii) All of us now have motivation to work harder with the examples of our senior scientists in front of us.

(iv) Personal contacts developed with leading workers and it created opportunities for seeking help and guidance for all times to come.

(v) A positive contribution was made by the developed countries for the betterment of the Third World countries.

(vi) Through exchange of knowledge and mental harmony, a positive contribution for betterment of mankind and promotion of world peace was made.

It is hoped that the high spirit of the workshop participants through this volume will spread to wider circles and fertilize mycorrhiza research throughout the tropics.

Author index

Subject index

An index of scientific names of vascular plants and fungi is given on page 266

adaptability 35-6, 42, 51, 56, 58, 64-5, 116
adverse sites 13, 43-6, 52, 55-61, 116, 217,
  242-3
afforestation 13, 78, 93, 96, 106, 121, 154,
  194, 251-3
agar 27, 29, 30, 33, 35, 45, 49, 50, 83
aggressiveness 52
agriculture 8, 9, 104, 139, 166, 197, 217,
  242, 252, 254
allelopathy 184, 254
ammonium 171, 195
Angola 137, 252
antagonism 51, 137
antibiotics 180, 218
appressorium 234, 237
arbuscle 145, 213, 235, 238, 245
Argentina 29-31, 48
Australia 3, 5, 31-3, 54, 66, 113-14, 130,
  219
Austria 26-9, 48, 53
avocado 197, 218, 247

bacteria 37, 53, 175, 179, 218
barley 216, 220, 222
basidiomycetes 14, 75, 96, 131, 175
basidiospores 19-25, 29, 58, 61, 138
  collection of 23, 24
  germination of 20, 24-5
biological control 247
borrow pit 35, 52, 59
bracken 118-20
Brazil 114, 128-9, 156, 197, 219

carbohydrates 29, 192
cassava 216, 224-5, 247
chlamydospores 20, 180
citrus 128, 213-14, 220-2, 224-5, 247, 254
clamp connections 130, 134, 150
climate 51, 73-4, 103, 110-11, 154, 166
clover 170, 183-4, 213, 219-20, 222, 231,
  254
coal spoil 35-6, 52, 55-7, 217
cocoa 128, 132, 197, 213, 232-7, 247
coconut 128, 247
coffee 184, 213, 225, 232, 247
colonization 15, 22, 29, 31, 37-40, 43-6,
  51, 55-6, 59, 64
competition 21, 36, 38, 43, 80, 165, 179,
  182-4, 214
contamination 25-7, 29, 33, 52, 74, 224

Costa Rica 130, 132, 156, 184, 197
cotton 128, 222, 247
cowpea 132, 184, 220, 222-4, 247-8
Cuba 156-62
cypress 8, 128

2,4 D 121-2
damping-off 31, 37, 99, 105
DBCP 245-6
deodar 104
dichotomy 96-7, 199
dipterocarps 108-9, 129, 146, 148-52, 252
disease 6, 137, 165, 180, 185, 217-18, 225,
  248, 254
drought 36, 45, 87

ecological adaptation, *see* adaptability
ectendomycorrhiza 5, 6, 19, 98, 115, 144,
  194
ectomycorrhiza 3-5, 8, 9, 13-66, 88, 121-2,
  144, 146, 148-50, 155, 165, 171, 184,
  194, 213, 218, 225, 251-2
ectomycorrhizal association 9, 13-14, 18,
  128-9, 133-8, 154, 232, 252
ectomycorrhizal development 14-16, 20,
  23, 26, 32-4, 40-1, 46, 54-9
ectomycorrhizal fungi 8, 13-20, 25, 30-1,
  64-6, 80-7, 92, 93-5, 108, 110-16,
  121-2, 130, 135, 137, 155, 166, 177-8,
  251-3
ectomycorrhizal inoculation 13-66, 90-5,
  251
ectomycorrhizal trees 3, 43, 111, 121
ectotrophic mycorrhiza, *see* ectomycorrhiza
endomycorrhiza 8, 9, 127, 146-7, 206-8,
  232, 246; *see also* VA mycorrhiza
endomycorrhizal association 8, 9, 111, 127-
  30
endomycorrhizal fungi 131, 166, 253
endotrophic mycorrhiza, *see* endomycorrhiza
endophyte 127, 192-3, 196, 206, 213-17,
  219, 223-5, 238-9
erosion 35, 58, 62, 217-18
eucalypts 14, 15, 19, 74, 90-2, 194
exotic trees 3, 7, 9, 13, 36, 65, 80, 91, 104,
  108, 121, 194, 196, 252
external mycelium 5, 144, 173, 213-14,
  225, 233-7

Index of scientific names

Vascular plants

Fungi